THE US 8th AIR FORCE AT WARTON 1942-1945
The World's Greatest Air Depot

THE US 8th AIR FORCE AT WARTON 1942-1945

The World's Greatest Air Depot

HARRY HOLMES

Motorbooks International
Publishers & Wholesalers

Dedication

This book is respectfully dedicated to all who served at Warton during World War Two and especially to those who made the ultimate sacrifice.

After this book was completed it was learned that Colonel Paul Jackson had passed away on 11 October 1997 aged 95. He was a true Officer and a Gentleman.

This edition first published in 1998 by Motorbooks International, Publishers & Wholesalers, 729 Prospect Avenue, PO Box 1, Osceola, WI 54020, USA.

© Harry Holmes 1998

Previously published by Airlife Publishing Ltd, Shrewsbury, England.

All rights reserved. With the exception of quoting brief passages for the purposes of review no part of this publication may be reproduced without prior written permission from the publisher.

Motorbooks International is a certified trademark, registered with the United States Patent Office.

The information in this book is true and complete to the best of our knowledge. All recommendations are made without any guarantee on the part of the author or publisher, who also disclaim any liability incurred in connection with the use of this data or specific details.

We recognize that some words, model names and designations, for example, mentioned herein are the property of the trademark holder. We use them for identification purposes only. This is not an official publication.

Motorbooks International books are also available at discounts in bulk quantity for industrial or sales-promotional use. For details write to Special Sales Manager at the publisher's address.

Library of Congress Cataloging-in-Publication Data available

ISBN 0-7603-0577-3

Printed and bound in England.

Contents

Acknowledgements 6

Introduction 7

Chapter 1   Air Depots – The Requirement 13

Chapter 2   Warton – The Early Days 19

Chapter 3   1944 – The Big Year 31

Chapter 4   The Saga of *Spare Parts* 47

Chapter 5   The Invasion – The Urgency Multiplies 53

Chapter 6   The Freckleton Disaster 75

Chapter 7   Returning to Normality 81

Chapter 8   Change of Command 93

Chapter 9   Victory in Sight 107

Chapter 10   V.E. Day – The Run Down Commences 123

Epilogue 134

Appendices

A   Warton Airfield – Layout and Location 135

B   Warton Airfield 141

C   Warton Air Depot (B.A.D. 2) – Complete Output 142

D   Section tasks undertaken by Maintenance Division 143

E   Units assigned to Warton 145

F   Aircraft types processed by Warton Air Depot 149

G   Example of B.A.D. 2 work processing – October 1944 150

H   Hangar 4 – Work in progress – October 1944 151

I   Aircraft visiting Warton 153

Glossary 158

Acknowledgements

To write one book to record the history of the Warton Air Depot, later designated B.A.D. 2, would be impossible as the unit's achievements during World War Two would fill volumes. However, this publication will attempt to represent the work of Warton from a slow start in 1942 to the magnificent output in production at the time the war ended.

B.A.D. 2 grew from a handful of Americans to an organisation of over 10,000. They did not fly in combat and only a small number were aviators, but their resourcefulness, loyalty and devotion to duty were unquestioned. Sacrifices were made, but the personnel did not receive any honour or glory although the tasks they completed contributed greatly to the success of the United States 8th and 9th Air Forces.

Much has been published about the combat units of the U.S.A.A.F. and I felt strongly that a history of B.A.D. 2 should take its place among the others. I have attempted to record the facts as faithfully as possible, which sometimes proved difficult as the U.S.A.F.'s historical records occasionally differ with those of the unit and, after fifty-odd years, individuals' memories start to fade and facts get mixed up with fiction. In every case of doubt I have reported the official version.

Many people have contributed to this book and I would like to give my sincere thanks to the following members of the B.A.D. 2 Association: Dave Mayor, founder of the association; Spence Thwaites; George Gosney, for his unstinting help; Ralph Scott; Orville Wrosch; Charlie Himes; Wally Woltemath; Pete Swank; Harry Goldsmith; Fay Brandis, who has been particularly helpful; Ray Dlouhy; Bob Tatro; Arthur Loscher; and Stan Ruggles. Fellow members who are no longer with us include – Jack Knight, Lee Maudlin, Ted Tryba, Paul Oberdorf and Edgar Stoke. Nellie Oberdorf, widow of Paul, and May Stoke, widow of Edgar, were also of great help. At the English end, I give my great thanks to: Arthur Talbot, current president of B.A.D. 2 Association; Aldon Ferguson who, as always, was a great source of help; Bob Fairclough and Ian Lawrenson of BAe Warton Heritage; Russell Brown, a local aviation historian; and last, but by no means least, Roger Freeman, a friend for over forty years, who knows more about the 8th Air Force than anybody! My thanks also go to Ian Webster of Photech of Oldham for getting the best out of some old negatives, and Peter Nield for his excellent plan of the airfield at Warton.

My 'fence hanging' days at wartime Warton will always remain an unforgettable highlight in my life and I hope that, in some small way, this book will help to repay some of the American kindness freely given both in those far-off days and today. I also hope that the remaining Wartoneers who read this book will look back with pride at their accomplishments, which played a major part in the air war over Europe.

HARRY HOLMES
MANCHESTER

Introduction

The name of Warton is still magical to me. It was where my love of aircraft really blossomed and it gives me much satisfaction to know that the site, as Headquarters of the Military Aircraft Division of British Aerospace, is still at the forefront of aviation technology just as it was in those days over fifty years ago.

In common with most wartime schoolboys I had an interest in aeroplanes and built the multitude of models issued by the now almost forgotten names of Skyleada, Airyda, Skyrova and other kit manufacturers. Up to 1943 I was content to do this, but then felt the desire to see real aircraft, and apart from the Spitfires, Mustang Is and Messerschmitt 109s with which the R.A.F. toured the country to raise funds for the war effort – plus passing glimpses of aeroplanes at Ringway and Burtonwood – I had seen very little at close quarters.

With my father otherwise engaged and my mother working, my grandmother became my constant companion. My school holidays were spent in and around Blackpool staying with friends and relatives, and the area was

The famous Blackpool Tower photographed in 1943. The town became a great favourite of many Americans, especially the turreted building which was the Palatine Hotel.

Above:
The P-38 provided much early work for Warton, but this aircraft type was later allocated to B.A.D. 1 at Burtonwood while the former specialised on the P-51 Mustang. The airfield's unique control tower is well illustrated in this shot.

Below:
Double parked! These P-51Ds awaiting delivery now have the invasion stripes painted on at Warton. The nearest aircraft, 44–63263, had a long career for, after serving with the 364th Fighter Group at Honington as 5Y-L of the 384th Squadron, it went on to fly for more than twenty years with the Royal Swedish Air Force as 26058.

always excellent for seeing aircraft with the many R.A.F. or Royal Navy aircraft from nearby stations usually making a beeline for the resort's famous tower or making low passes over the sands. Squires Gate aerodrome was always worth a visit and many a happy hour was spent just watching the aircraft operating from there.

However, in the middle of 1943 I began to notice that uniforms were not only R.A.F. blue but American khaki, and my aunt told me that the U.S.A.A.F. had taken over Warton aerodrome near Lytham St Annes, just a few miles away. I don't think this news stirred me very much, but one day while on the Golden Mile I heard the sound of an aircraft which was new to me. I looked up to see the beautiful lines of a Lockheed P-38 Lightning which came in from the sea to circle around Blackpool Tower before heading off down the coast. Although I had never seen one, its unique twin-boom shape was easily recognisable as it had been well illustrated in the many aircraft publications of the time. This was the handsomest aeroplane I had ever seen and I wanted to see more of it. As it was American it was a near certainty that the P-38 had come from Warton and next morning my grandmother and I set out by bus to find the airfield.

As part of my education my grandmother encouraged me to keep a diary, and after a few visits to Warton my daily events began to take on a new meaning. I started to make notes on aircraft colours and code letters as an aid to modelling, but sadly I did not record many serial numbers as they were usually too small to see from outside the airfield, and even though I did own a small telescope it was not wise to be seen looking through one in those days, especially at military installations!

After my first few visits I began to go alone, and in my notebook the first recorded visit was 13 September 1943 when I noted all the details of a B-17 Fortress which was parked near the fence, as were four brand-new P-38s and an R.A.F. Stirling. A number of other types were on the airfield or in the circuit. The B-17 belonged to the 92nd Bomb Group at Podington and the Stirling was owned by 149 Squadron, then stationed at Lakenheath. I know these things now, but at that time I had absolutely no idea of the code/unit combinations or where any of the visiting aircraft came from. I had many friends who were modellers, but we really did not have any knowledge of the significance of the codes, and I suppose that was the general idea. After the war, when I started to correspond with other enthusiasts, the combinations began to be established, although there has never been an official list of the allocations and even today there are still mysteries. Regarding the U.S. 8th and 9th Air Forces, as Warton had no unit markings of its own it was hard to follow any pattern especially with the variety of aircraft arriving at the station carrying different unit markings. Such was my naivety during the early visits I thought that Warton was an operational airfield which, in the evenings when I was not around, launched missions against Germany!

It was, of course, Base Air Depot No. 2, Station 582 of the U.S. 8th Air Force which later became known as 'The World's Greatest Air Depot' after setting many production records. B.A.D. 2, when fully established, had over 10,000 men working around the clock on every type of overhaul, modification and general maintenance on aircraft and equipment. The first Americans arrived at Warton in July 1942 and from mid-1943, when production got into its stride, until the war's end Warton processed over 10,000 aircraft including an incredible 4,372 P-51 Mustangs and 2,894 B-24 Liberators, with an assortment of other U.S. types making up the rest. In fact, every type of American aircraft used in the European Theatre of Operations (E.T.O.) visited B.A.D. 2 at some time. The unit also completely overhauled in excess of 6,000 aero-engines with more than half this number being the Packard-built Rolls-Royce Merlin.

One of the many records established was set in March 1945 when a total of eight hundred and fifty-three aircraft passed through the depot to be ferried to combat units, and I well remember seeing lines of P-51s, B-24s and A-26s which I thought were based at Warton as I did not realise they were constantly changing. There were over 45,000 aircraft movements from Warton in the time in which B.A.D. 2 operated, and 1944 alone saw 22,000, so for the enthusiast it was paradise as aircraft were constantly taking off or landing and it was not unusual to see five different types in the circuit at the same time. In the autumn of 1944 the airfield was so full of aircraft that two of the three runways had to be used for parking, and at one period over 800 were on the field.

Warton had its share of accidents, but one tended not to hear any details of these although the disaster of the B-24 hitting the school in nearby Freckleton was well-known. These incidents will be covered later in the book, but when I visited the airfield on 1 August 1944 I had no idea of two fatal P-51 crashes a few weeks earlier. My interest that day was in a beautiful yellow-nosed P-51 Mustang which was parked not too far from the fence. The note in my book shows it having the name 'Louis', but I found out years later that it was actually called *Lou IV* and was the personal aircraft of Colonel Thomas J.J. Christian Jr, who was commander of the 361st Fighter Group. He was to lose his life in this machine when it was shot down over Germany less than two weeks later.

Despite the sadness of things like this there were always some lighter moments, including the time I got a clip on the ear from the local bobby who told me that if I wanted to collect numbers I should know better than to copy down details of aeroplanes in wartime. After the lecture he relieved me of my notebook and recommended that I visit the railway to collect numbers! Luckily, I had used the pocket book for quick notes which I transferred into a large book at home, but after that encounter I used pieces of paper in case I was caught again.

During my numerous visits I got to know many of the Americans based at Warton, and especially some of the military policemen who patrolled the various gates. Many shared their lunches, candy, gum and fruit with me with the kindness and generosity for which their nation is famous. Luckily, I am able to maintain many friendships through the

Seen from the waist gun position of another Liberator, this B-24H 42–94894 had a complete respray after overhaul although its original shark's mouth nose art was too good to paint over.

Introduction

*The Mustang **Lou IV** was the personal aircraft of Colonel Thomas J.J. Christian Jr, the C.O. of the 361st Fighter Group, seen by the author on 1 August 1944. Less than two weeks after its visit to Warton, the P-51D was shot down with Colonel Christian losing his life.*

B.A.D. 2 Association.

One meeting I will never forget took place in September 1944 when my faithful grandmother and I were sitting on some steps which led from the promenade, near Blackpool Tower, down to the sands. It was a lovely evening and we got into a conversation with a young American pilot who was feeling a little lonely, and we chatted for a while with the usual family photographs being shown. I told him of my interest and he said that he had come to collect a B-24 Liberator from Warton the following day. I told him where I usually sat and he promised that if I was there in the morning he would rock his wings for me on take-off. Needless to say, at 7.30 a.m. next morning, 9 September, I was in position on a bright but damp morning and I did not have to wait long until a B-24 started to taxi out for take-off. My excitement was shortlived for as it flew by and climbed away there was no sign of wing movement. Choking with disappointment I settled down to watch the other activity around the field and as the movements and the temperature began to increase I felt a little better. It was about ten o'clock when a shiny Liberator lumbered down the runway on take-off, and as it went by I stared in amazement as there seemed to be someone waving to me from every position and the tears really welled up in my eyes as the wings began to rock. The bomber circled the airfield and then headed off to the southeast, no doubt to the Second Air Division. I always wondered if that young pilot survived the war, and it will always be one of my cherished memories to think that it was done just for me.

For many members of the B.A.D. 2 Association their tour of duty at Warton was the most exciting time in their lives and I am still amazed at the enthusiasm to visit the base 'just one more time'. Knowing how things were, I can fully understand their feelings and I would give anything to see a Mustang rifling down Warton's main runway at 'zero' feet before peeling up for a landing, or to see a B-24 taking off looking ridiculous beside those beautiful P-51s. Ah well!

It was my visits to Warton and my love of

The World's Greatest Air Depot

aeroplanes which really determined my future, and as I was employed by British Aerospace I was privileged to visit Warton for various meetings and even to fly from its runways. Even now, after retirement, I cannot resist the temptation of a Sunday drive which just happens to pass by Warton where I can motor round to my old spotting position, close my eyes, and drift back to those magic days of the Eighth Air Force. The world will never see its like again!

H.H

Jack Knight test flies this beautifully clean Mustang above the Lancashire overcast in May 1944.

Chapter 1

Air Depots – The Requirement

In October 1939 Lord Beaverbrook, Minister of Aircraft Production, recognised the requirement for maintenance and repair facilities for the ever-increasing number of aircraft being supplied by the United States under the Lend-Lease programme. Beaverbrook had insisted that everything connected with the production of aircraft should come under his ministry's control so he formed the Directorate General of Aircraft Production (Factories) to establish every facility required by the M.A.P.

The American aircraft were arriving in large numbers, but the R.A.F. was handicapped by a lack of personnel familiar with these types and their equipment, so the U.S. factories sent their representatives to initiate the British into the mysteries of American handbooks. It was also planned that a number of specialised sites be established to undertake the task. However, it was not until the spring of 1941 that these plans got underway, and they now incorporated the requirement of bases to be used by the U.S. Army Air Force.

On 19 May 1941 a team of U.S.A.A.F. observers under the command of Major-General James H. Chaney arrived at R.A.F. Northolt for a series of meetings in London and a tour of possible sites for American bases in the United Kingdom. The following month Brigadier-General Ralph Royce visited facilities in the Middle East where the R.A.F. were operating American-built aircraft, and this tour was followed, in August, by a similar visit by Major-General George H. Brett. In a joint report both Generals recommended that greater control over U.S. personnel and installations would lead to greater efficiency.

In October 1941 Lord Beaverbrook was requested by the U.S.A.A.F. to recommend a number of sites for possible use as an air depot. The terms of reference were limited by geographical areas and the task was undertaken by Air Vice-Marshal Weedon, who had been seconded to M.A.P. from the R.A.F., and a civilian engineering expert, Frank D. Thomas. After much travelling by air and numerous meetings, four sites were recommended: Langford Lodge near Loch Neagh in Northern Ireland; Warton on the Lancashire coast; Little Staughton near Bedford; and Burtonwood near Warrington, also in Lancashire, which, although not fully developed, was already engaged in the overhaul and repair of American-built airframes and engines. A visit by General Brett soon followed and after an inspection of the proposed sites he agreed with the recommendations of the M.A.P. team. The United States entered the war in December 1941, and by early January 1942 General Brett's recommendations were actioned, and in March a detailed agreement was reached by the U.S.A.A.F. and British authorities for the development of these sites.

Also in January 1942, Frank Thomas had been instructed by the D.G.A.P.(F.) that he was to take charge of the design and construction of these sites with each comprising a technical site with large hangar and storage space, personnel accommodation, a hospital and an airfield with three runways. The facilities were to be operational by the year's end, but the U.S.A.A.F. informed M.A.P. that Warton was to be in use by the end of October. The contractors were chosen for their previous record of efficiency and not by competitive tendering, and work commenced in March 1942.

13

The World's Greatest Air Depot

A Luftwaffe reconnaissance photograph of Warton dated 29 September 1940 when the area had been earmarked for the construction of a fighter station for the R.A.F. The airfield's outline, which had been inked in at a later date, was reasonably accurate as the photograph was re-issued for information in January 1943.

The duty of VIIIth A.F.S.C.'s Mobile Repair Units was to recover aircraft from the location of the crash-landing, and this shot shows that it was not always an easy task. The large recovery vehicle appears to have toppled over on to its side while trying to rescue this 356th Fighter Group P-47 Thunderbolt.

General Ira C. Eaker inspected the Burtonwood installation in April 1942, and acting upon Eaker's information General H.H. Arnold sought to secure its transfer to American control. This plan called for existing British technical staff to continue in service until U.S. technicians became available, and for the centralisation at Burtonwood of the supply and repair of all American-built aircraft. This would be the inauguration of a policy already agreed which gave the U.S.A.A.F. responsibility for the supply and maintenance, excluding modifications, of all U.S. aircraft operating from the United Kingdom. The urgency came from General Arnold's revelation that current plans proposed to have 1,000 American aircraft operating from bases in the U.K. by August 1942, and 3,500 by mid-1943. Until Warton and Langford Lodge were fully completed at the end of 1942, Burtonwood would have to serve instead.

On 23 May 1942 General Chaney and the ministry finalised a detailed agreement to transfer Burtonwood to the exclusive control of the Americans following an interval of joint operation. This joint control came at the end of June under the supervision of the VIIIth Air Force Service Command. With the shortage of skilled military personnel it was arranged that civilian technicians would be transferred from U.S.A.A.F. air depots in the United States to serve as a Civil Service Detachment at Burtonwood. The VIIIth A.F.S.C. was, of course, part of the 8th Air Force which had been activated at Savannah, Georgia on 28 January 1942 as part of a plan known as 'Gymnast' which included landings in North Africa, but this and an advanced version were changed and the involvement of the new 8th Air Force was not required. As there had been a requirement to establish the U.S.A.A.F. in the United Kingdom the 8th was selected for this task on 8 April 1942.

As construction at Warton and Langford Lodge fell behind schedule Burtonwood had to go it alone, but it was later established that even when the facility in Northern Ireland became operational its location would limit its usefulness to the combat groups. The delay in the development of the air depots forced the groups already operational to carry out their own heavy maintenance and repairs, and it was the site at Little Staughton which was given the highest construction priority as it was the nearest to the VIIIth Bomber Command's operating area. Under the direction of the A.F.S.C. a number of mobile repair units were activated to assist the combat bases and these small sections commenced operations in September 1942. Each section was equipped with a one-ton truck and a jeep plus two small trailers, all fitted out with the required tools and supplies to complete on-the-spot repairs of battle-damaged aircraft. The main task was to repair

The World's Greatest Air Depot

crash-landed aircraft at the location of landing and, if possible, fly them out to a suitable base for more permanent repairs. The first mobile repair unit received its full equipment by December and began heavy operations by New Year's Day 1943. Eventually more than fifty of these units were established, and by June 1943 they had repaired almost two hundred aircraft. During the last half of the year they returned an average of sixty-eight aircraft per month to their groups, proving themselves to be a valuable part of the 8th Air Force.

Warton, destined to share the bulk of the heavy maintenance with Burtonwood, was working at only ten per cent of its capacity by the end of June 1943 and it was suggested that Lockheed Overseas Corporation, manned by American civilians, should take over. However, the U.S.A.A.F. rejected this proposal as it rigidly stuck to its plan that all depots would be operated by military personnel. Langford Lodge started working on 19 November 1942 with the Lockheed Overseas Corporation which operated the base on contract to the U.S. War Department under the supervision of the VIIIth A.F.S.C. Nearly all types operated in the European Theatre of Operations (E.T.O.) had to be processed, but the inaccessibility of the unit from the combat bases in eastern England minimised its value as a heavy repair and assembly depot, and only flyable aeroplanes were taken in for repair. As Warton and Burtonwood were to take over this task fully, Langford Lodge devoted its efforts to modifications and engineering research.

Bomber aircraft were flown into Britain from the United States, but fighters arrived by ship at the ports of Liverpool or Glasgow, their bodies cocooned minus propeller, ailerons, wingtips and tail assembly in the case of the P-47s and P-51s, while the P-38s and later the P-61s arrived without their outer wings.

The new P-51D version of the Mustang commenced arriving at Liverpool at the end of May 1944, and this convoy is passing the Plaza cinema in Allerton Road en route to Speke aerodrome for assembly. The aircraft were then flown to Warton for inspection and any required modifications.

Air Depots – The Requirement

The construction of Nissen huts as accommodation continued apace as shown on Site 9, but the steady influx of personnel was faster than the construction work.

The smaller fighters were towed on low-loaders through the streets of Liverpool from the docks to Speke aerodrome while the twin-engined machines used their own undercarriage. The Lockheed Corporation at Speke assembled the aircraft and the P-51s were flown to Warton, the P-47s and P-38s flying the short distance to Burtonwood. The Glasgow arrivals were completed at Renfrew before flying on to their respective depots.

The joint Anglo-American operations at Burtonwood came to an end in October 1943 when the base was completely militarised as airmen replaced both British and American civilians, with the latter returning to the United States in the winter of 1943–4. The Lockheed Overseas Corporation contract at Langford Lodge was also cancelled as of 3 July 1943. The main build-up of personnel was to be at Warton and Burtonwood, and eventually this figure totalled over 25,000 which included women soldiers and nurses.

The now fully operational air depots were known as Base Air Depots (B.A.D.) No. 1 Burtonwood, No. 2 Warton and No. 3 Langford Lodge, and, for the former two bases, assembly line methods were introduced which helped in the utilisation of the large number of unskilled men who were attached to these units. In May 1942 a committee under the leadership of Major-General Follett Bradley had to establish a plan to explore completely the possibilities of operating, maintaining and supplying approximately 7,000 aircraft in the United Kingdom supported by a maximum of 500,000 U.S.A.A.F. personnel. The report became known as the Bradley Plan with one of the recommendations being to specialise the technicians in certain aircraft types and rationalise a number of the smaller units, amalgamating them with sections having reasonably similar functions and incorporating those into large divisions, thereby dispensing with a great deal of administration work.

Both Warton and Burtonwood utilised the

The line-up of P-47 Thunderbolts at Warton with just one Mustang belied the fact that the overhaul and repair of the heavy fighter had already been allocated to Burtonwood.

By the end of 1943 B-24 Liberators were starting to arrive in increasing numbers as B.A.D. 2 was to specialise on this type of bomber. This aircraft was the B-24D 42–63790.

Bradley Plan with units being allocated to Military Administration, Maintenance or Supply Divisions. The Maintenance Division, which included Flight Test, was by far the largest division at both facilities. The smaller Strategic Air Depots (S.A.D.) were not expanded greatly as it was recognised that B.A.D. 1 and B.A.D. 2 would share the load. However, it was also ordered that the smaller liaison-type aircraft which arrived in large crates could be transported directly to the S.A.D.s for assembly instead of to Speke or Renfrew, in order that Burtonwood and Warton could concentrate on the processing of the combat aircraft which arrived by sea. The modification of aircraft began to absorb an increasing number of hours in the maintenance effort from the summer of 1943, and the arrival of new aircraft increased the load further. Before the end of 1943 both Burtonwood and Warton were modifying B-26s and C-47s alongside the heavy bombers and fighters already underway.

The need for specialisation was no more apparent than in the engine overhaul sections at the air depots, and in December 1943 all radial engines for types such as the P-47, B-17, B-24, B-26 and C-47 were ordered to be processed at Burtonwood while the in-line engines used on the P-38 and P-51 were assigned to Warton, with this work commencing in January 1944. Langford Lodge, meanwhile, manufactured engine kits and repaired electrical propellers.

As previously noted, the demands for modifications were continually increasing as the air battle over Europe intensified and lessons were learned. It was not only tactics which were being updated, but the requirements for additional firepower, armour plating, better vision and a host of other improvements. As the expansion programme increased and the rationalisation was introduced, Burtonwood would take the lion's share of B-17s, P-38s and P-47s while Warton would be solely responsible for B-24s and P-51s, but would also take P-47s as required. Langford Lodge was also brought into the equation to modify some B-17s and P-38s. The Ninth Air Force, which was re-formed in England in October 1943 after serving in North Africa, would modify many of its own aircraft at advanced bases, but the pressures of the build-up and the forthcoming invasion would force them to allocate A-20s, B-26s and C-47s to the Base Air Depots, adding to their already great burden. Almost at the same time, the IXth A.F.S.C., which had responsibility for assembling liaison aircraft such as the L-4 and L-5, eventually had to pass this work to the B.A.D.s.

The work of the Base Air Depots continued until after the war in Europe was over and although it was the combat flyers who achieved the fame, it would not have been possible without the dedication of the men at the B.A.D.s.

Chapter 2

Warton – The Early Days

It was on 21 October 1938 that the Air Ministry announced an interest in Grange Farm, the Preston Corporation Sewage Farm located near the small village of Freckleton, as the possible site for an aerodrome. Freckleton was a sleepy village on the Lancashire coast to the north of the River Ribble and the prospect of an airfield was good news for the local council as unemployment in the area was higher than the national average and it was felt that work could be found for many men. It was particularly pleasing to the Fylde Rural Council who had directed the attention of the Air Ministry to the possibilities of Freckleton Marsh when the ministry was searching for an aerodrome site in Lancashire at the end of 1936. The council's overtures had been politely received and filed for consideration.

Just one month later, on 18 November, the council's jubilation was shattered by a letter from the Air Minister, Sir Kingsley Wood, stating that Grange Farm was unsuitable for the ministry's purposes. It was disclosed that their technical advisers reported that the ground at the site was unsuitable for the erection of buildings for an aircraft factory and the adjacent airfield. The letter, addressed to Preston Council, also stated that the building of an aerodrome at Samlesbury would be changed with a factory; originally proposed for Vickers-Armstrong, being allocated to the English Electric Company for the production of the Handley Page Hampden medium bomber. Fylde Rural Council strongly objected, stating that there were certainly no technical difficulties in adopting the marsh as an aerodrome, but the decision had been made.

After the outbreak of the Second World War it was obvious that the need for aerodromes would take on a greater urgency and sites previously rejected were reassessed. The Grange Farm area came under scrutiny once more, but by moving a further half a mile along the coast towards the village of Warton, the marshland there was much more suitable for an airfield. In 1940 work commenced on the site to construct an airfield for the R.A.F. to be used as a satellite for nearby Squires Gate, where detachments of a number of fighter and bomber squadrons had already been rotated. Once again the decision for an airfield was welcomed by the council although a number of the local poultry farmers greeted the news with anger as a number of farms would have to make way for the aerodrome.

After America's entry into the war, decisions were soon made on the requirement of air bases in the United Kingdom, and by January 1942 the plan for Warton as an air depot site was already in place. By March Frank Thomas of the Ministry of Aircraft Production had confirmed Sir Alfred McAlpine and Son Limited as principal contractors. Two runways were under construction at that time, but there were few buildings on the site and it was not until June 1942 that the real construction programme was started. By the end of that year all the planned work, apart from the living accommodation on Site 13, had been completed. One sad note for the locals came not only with the loss of a number of farms, but with the demolition of several residences built by retired sea captains and a favourite hostelry known as the 'Guides House' which was situated on the banks of the River Ribble.

It was 18 August 1942 when the first contingent of United States Army Air Force

The World's Greatest Air Depot

One of the early fighter types at Warton was the Lockheed P-38 Lightning, but this example had a rather inauspicious arrival on 26 November 1942. The aircraft, 43–2087, was a P-38F taken over from a British contract and it seemed unusual to see a '43 serialled aeroplane in 1942.

Another view of the belly-landed Lightning.

Warton – The Early Days

A rare type in Britain, this Bell P-39M Airacrobra, 42–4833, ended its days at Warton on 3 February 1943. The pilot, Second Lieutenant Potts, was able to escape from the aircraft before it was destroyed by fire.

*One of the first U.S.A.A.F. aircraft to visit Warton was the B-17E 41–9022 named **Alabama Exterminator II** of the 97th Bomb Group.*

personnel arrived at Lytham railway station, which was just four miles from Warton. These men would be the first of many, for at its peak the base was home to over 10,000 personnel and the whole Lancashire coast would soon begin to feel the impact of the Americans' arrival in the area. The four officers and ninety-two enlisted men were greeted by Wing Commander Styles and Squadron Leader Hillier from R.A.F. Warton, and the following day a further thirty-eight officers and six hundred and forty-three enlisted men arrived. This pioneering group was under the command of Major John Shuttles.

The Americans arriving at Warton would soon be working hard to get the base established, but the possibility of playing hard was not lost on them as they were soon able to sample the delights of Blackpool, the famous holiday resort, which was compared by many to their own Coney Island or Atlantic City. Blackpool was no stranger to the American military as during the First World War eight hundred troops of the U.S. Base Hospital Medical Corps were stationed in the resort from May 1917.

On 5 September 1942 the Warton Air Depot was established by order of the VIIIth Air Force Service Command. On 9 October Colonel Ira A. Rader assumed command of the depot to become Warton's first Commanding Officer.

Aircraft were already paying regular visits to Warton, but excitement came on 26 November 1942 when fire crews and ambulances were alerted as an approaching Lockheed P-38 Lightning had radioed a distress call. The aircraft touched down on the main runway but skidded off into the grass ripping off the undercarriage and ending up on its belly; happily, the pilot climbed out unhurt. Besides being Warton's first American accident statistic, the aircraft was most interesting as it was P-38F, 43–2087, a 1943 serial number taken over from an ex-British contract.

Originally planned as the 402nd Air Depot, Warton's main task was the overhaul, modification and repair of U.S. aircraft operated in the E.T.O. by America or her allies. Aeroplanes arriving from the United States would have to be adapted to meet the demanding requirements of the European Theatre of Operations and the ever-changing strategic and tactical needs would constantly require new or updated equipment to be installed in combat aircraft. Battle-damaged aircraft would have to be repaired with perhaps whole sections having to be rebuilt or changed. Those aircraft which were beyond repair would go to a specialised salvage section which would relieve the machine of every usable part for storage or immediate incorporation into another aircraft. These requirements called for every type of technical skill on a variety of aircraft models. Also required would be a means of flying aircraft from other locations into Warton and, after processing, delivering them to their own bases or other airfields. For this operation the Warton Air Depot would later be assigned the 310th Ferrying Squadron.

The growth of the station was slow and by the end of 1942 there were only four units officially assigned with the 977th Military Police Company, the 162nd Ordnance Company, the 407th Truck Company and the 7th Air Depot Group being the first at Warton. In January 1943, although still badly needed at Warton, a detachment from the 7th Air Depot Group was sent to Langford Lodge in Northern Ireland but returned almost immediately as Warton's expansion was given a higher priority.

On 3 February 1943 the emergency services were called out as an arriving aircraft reported trouble with the undercarriage. The aircraft made a perfect belly-landing but was destroyed by the ensuing fire with the pilot, Lieutenant Potts, walking away from the machine with minor injuries. Once again, it was an interesting incident as the aircraft was a P-39M Airacobra, 42–4833, a rare type in the United Kingdom.

On 12 February 1943 Colonel Rader was replaced as Commanding Officer by Colonel A.S. Albro. Colonel Rader had been trying to establish a working system for the depot, but

This B-24D Liberator at Warton for overhaul had a cover over its Norden bombsight as, in the early days, the instrument was highly classified and an armed guard was always present. The security was relaxed when it was found that the Germans had full details of the bombsight.

The World's Greatest Air Depot

The Wings for Victory parade in Blackpool on 19 April 1943 was the first chance the Americans had of taking part in any review. The troops made a good impression which was quite satisfying to Colonel W.D. Lucy, their Commanding Officer, especially as some had only arrived at the base two days earlier.

The hand-over ceremony took place on 17 July 1943 with a smart march-past of the reviewing stand erected outside Hangar 4. The R.A.F. band provided the musical accompaniment.

The parade seen from the newly erected American-type control tower with a variety of aircraft providing the backdrop. Outside the Flight Test hangars are a UC-61 Forwarder, a P-39 Airacobra, an R.A.F. Tiger Moth, and Warton's own B-26 Marauder **Demon Deacon**.

Lieutenant-General Henry J.F. Miller, Commanding General of the 8th A.F.S.C., accepts Warton on behalf of the United States government from Air Vice-Marshal P.D. Robertson, R.A.F.

things did not seem to be moving fast enough for A.F.S.C. Headquarters. At about the time of Colonel Albro's arrival the facility's name was changed from Warton Air Depot to A.A.F. Station 582 for security reasons. Colonel Albro took steps to speed up the work process, but after just two months he too was on his way. His unexpected departure left an empty chair in the C.O.'s office and it was left to Major Creed de Vazeille, the Base Construction Officer, to take charge until a senior officer could be found to take command. Just two days later, on 18 April 1943, Colonel W.D. Lucy arrived from A.F.S.C. Headquarters, but he was already destined for higher things and he too moved out after only three weeks. He later became Deputy Commander of the forthcoming Base Air Depot Area. On 6 May 1943 Colonel Charles W. Steinmetz was flown in and immediately proved to be a strict disciplinarian. Some of his methods were not well received by some of Warton's personnel, but under his command the depot did start to get results, although this was assumed to be by natural evolution than by other causes.

However, the succession of commanding officers continued as less than two months later Colonel Steinmetz was transferred to another station, on 30 June 1943. The following morning Colonel John O'Hara Jr stepped into the breach, and just over two weeks later on 17 July 1943 he was by the side of Major-General Henry J.F. Miller, Commanding General VIIIth A.F.S.C., on the reviewing stand as Warton was handed over to the Americans by the R.A.F. The A.O.C. of 17 Group, Air Vice-Marshal P.D. Robertson, R.A.F., was on hand to make the presentation and expressed his pleasure at seeing such a smart turn-out by Warton's troops. For many it was their first real parade since their training days in the United States, but as soon as the ceremony had ended little time was wasted before most of them were back in overalls.

Under Colonel O'Hara's leadership the depot started to reach its early targets as personnel were getting experienced with more facilities and material becoming available. Most of the time had been spent setting up the depot as a production facility and as everything was starting to fit into place, things began to happen.

Although processing had commenced in April 1943 it was not until July that the first statistical report reached the Commanding Officer. Aircraft were passing through the depot, but these were arriving for inspection and clearance on to the assigned units. These aeroplanes did not feature in production records and it was August before any actually appeared, as three B-17s had been processed and delivered.

After numerous odd jobs, Warton's first real task was the installation of external life-raft release switches, modification of radio and intercom equipment and the enlargement of ammunition boxes on B-17s. Because so many sections were involved in this work it soon became apparent that drastic rationalisation would be needed.

As production was destined to expand rapidly as the U.S.A.A.F. build-up continued, the busiest unit at Warton was bound to be Flying Control. A small detachment of men had arrived from the 7th Air Depot Group,

The R.A.F.-type watch office was used for air traffic control until the Americans erected their own style of tower, this time on stilts as used in the United States.

Another arrival which proved unfortunate occurred in May 1943 when the B-17F 42–30071 had a landing gear failure. The aircraft was repaired at Warton and went on to join the 100th Bomb Group at Thorpe Abbotts in Norfolk, becoming **Skipper II** *with the 418th Squadron.*

and after getting some experience at an R.A.F. station in southern England a full-time Flying Control section was set up in February 1943. Their equipment comprised of two Very pistols and an Aldis lamp plus one radio set, but as the volume of flying increased so did the much-needed supplies of communications materials. The addition of runway lights was also a great help. The first real excitement for this new section came early in May 1943 when a flight of B-17s arrived over the field. All went well until one pilot reported that he could not lower his wheels and he was advised to circle the airfield until arrangements could be made for an emergency landing. Controller Edgar Stoke drove a jeep down the airfield laying out coloured streamers on the grass to guide the bomber into the safest part of the field. The pilot then made a dry run over the grass, circled the airfield, made a good approach and then promptly set the Fortress down on its belly in the middle of the main runway.

On 2 April 1943 Flying Officer E. Gore-Rees, R.A.F., arrived to provide experienced air traffic cover, but he was already ill and two days later he was rushed to hospital at R.A.F.

Edgar Stoke helped to establish the Air Traffic Control section at Warton, but when he did so he could never have envisaged the large number of aircraft which would be handled before the war's end.

Warton – The Early Days

A Chevrolet truck used as a runway control vehicle had an observation dome fashioned from the top turret of a B-24 Liberator.

Station Headquarters building as it appeared soon after the handing-over ceremony in July 1943.

Weeton and did not return to Warton until two weeks later. The current controllers, all enlisted men, had done an excellent job, but as the traffic was increasing it was felt that more experienced people were needed, and the first American officer controller arrived on 21 April. He was Second Lieutenant John C. Taylor, and just over one week later another American, Second Lieutenant James H. Shaw,

The arrival of so many before the huts could be completed meant that the incoming troops had to endure the winter of 1943–4 in 'Tent City', and the cold, damp weather with the inevitable mud made conditions extremely hard, especially for those from the 'sunshine states'.

joined the team.

After much hard work in establishing a reliable air traffic system at Warton, good airfield discipline was taught to all and new navigation aids began to appear with airfield obstructions being removed, controlled or marked. By the beginning of August an Airfield Control Van was operating and runway numbers were painted on each landing strip. In recognition of their hard work and dedication Gore-Rees was promoted to Flight Lieutenant and both Taylor and Shaw went up one rank, all with effect from 9 September 1943. On 25 September the Flying Control section came under one command with Gore-Rees being replaced by Second Lieutenant John G. Shortall, who reported for duty after being trained at R.A.F. St Eval in Cornwall. Gore-Rees stayed on acting as a liaison officer for the station, but on 9 December 1943 he reluctantly left Warton to start all over again, this time training American controllers for the 9th Air Force.

Personnel continued to arrive at Warton at an amazing rate, and due to the shortage of accommodation a large tented area known as 'Tent City' was established; unfortunately, many of the troops had to spend the winter of

1943–4 under canvas. One of the arrivals in August 1943 was Major Charles W. Himes with his 314th Depot Repair Squadron, and it was he who later became Warton's chief test pilot and a popular figure around the base.

On 21 October 1943, A.A.F. Station 582 became known as Base Air Depot No. 2, becoming part of the newly created Base Air Depot Area Command (B.A.D.A.). Changes also came on 29 October when Colonel John G. Moore was appointed Commanding Officer of B.A.D. 2, and in later months his leadership contributed greatly to the steady increase in production. Colonel Moore's motto – 'It can be done' – was adopted by B.A.D. 2 and this proved to be true as the Depot went on to break many production records in terms of aircraft processed and delivered to active units.

Around that time large-scale changes were underway and the rationalisation needed was coming into effect. Individual units were inactivated with personnel and equipment absorbed into three newly created divisions. The largest of these was the Maintenance Division with the Supply and Administration Divisions making up the other two. On 1 November 1943 the divisions were officially transferred to B.A.D. 2 after having a 'provisional' status for five days. The U.S.A.A.F.'s Bradley Plan was beginning to take effect. Warton's whole mission concept could really be said to be encompassed by the Maintenance Division. Under this division came the Aircraft Section, Manufacture and Repair, Production Inspection, Production Planning, Accessories and Miscellaneous, and the Engine Repair Section. The Aircraft Section, of course, included Flight Test.

In October 1943, due to the increasing number of aircraft using Warton, it was decreed that separate call-signs be used for station aircraft and those of the Warton-based 87th Air Transport Squadron. This Squadron was part of the 27th Air Transport Group which was based at Heston, but the 87th also had detachments at Burtonwood and Langford Lodge. The Squadron's task was to fly transport operations and ferry aircraft between the depots and combat units. This role was carried out until November 1943 when the 2920th Ferrying Squadron (Provisional) was activated, but this unit lasted only three days before it was deactivated and replaced by the 310th Ferrying Squadron. The result was much more

Three of Warton's test pilots pose with a new type arriving at the depot, the P-51 Mustang. Lloyd Bingham, left, who was lost soon afterwards when his P-51 dived into the sea off Blackpool, is seen here with Jack Knight, centre, and Major Charles Himes.

This evocative shot looking east from the tower shows a growing number of aircraft in November 1943. The R.A.F. were always pleased to visit any American base and Warton was no exception with a Tiger Moth, Anson, Beaufighter and Hurricane on the ramp.

On the western side of the ramp are Flight Test Hangars 6 and 7, with an Oxford and Dominie in U.S. markings and a Mustang in R.A.F. colours.

In December Blackpool's promenade is always deserted and 1943 was no different as this photograph taken by Jack Knight from a P-51 will testify. To the right of the tower is Central Station where, two years earlier, a Blackburn Botha had crashed after colliding with a Boulton Paul Defiant

efficient with the 310th taking over all ferrying duties while the 87th was free to continue its air transport activities. In November 1943 1,216 aircraft were delivered from Warton, and although the majority of them had been new arrivals from the United States only cleared after inspection, one hundred and nineteen had been processed by B.A.D. 2.

Also in October a new aircraft type arrived at Warton with the appearance of the first P-51B Mustangs, and work commenced immediately to familiarise personnel with the new fighter. Almost straight away problems started to arise with a shortage of radiator gaskets, but this trouble was relieved by B.A.D. 2 ingenuity when it was found that a rubber composite product named Sheerline could be adapted for the purpose. The substitute was cleared by the aircraft's manufacturers, North American Aviation.

Another large contingent of personnel arrived in November 1943, and by the end of that month the Warton head count had reached 8,546 Americans.

On 6 November Warton suffered the loss of one of its leading technical specialists when Major George R. King was posted missing after the aircraft in which he had hitched a ride had disappeared. The B-26 Marauder 40–1372 was flying to Toome in Northern Ireland, a station used for training combat crews for the IXth Bomber Command, when the aircraft was lost over the Irish Sea. No trace of the machine was found.

The great increase in production was not

The ingenuity of the Warton technicians was never ending as new methods were devised for reducing man hours. One such came from Master Sergeant Fred Covert who designed a method of removing the undercarriage leg of a B-24. Normally taking five men working on a makeshift platform, as the photograph illustrates, Covert reduced the task to two men completing the job in half the time with considerably more safety.

The Aircraft Movement Section in December 1943 with Fay Brandis and Joseph Stenglein of the 310th Ferrying Squadron studying the operations board.

just down to the Commanding Officer and his officers: many of the B.A.D. 2 personnel came up with ideas to speed up jobs and lessen the time spent completing the task. For instance, when a B-24 Liberator arrived, one of the modifications required the removal of the landing gear struts for a bearing race change. The mechanics had to climb on to a platform under the wing to remove the massive strut by hand with each of the legs weighing over one quarter of a ton. Master Sergeant Fred Covert designed a landing gear strut remover jack which enabled the five-man job to be done by just two with much less effort and considerably more safety.

With the rationalisation the Engine Repair Section would specialise on in-line engines while the radials would be allocated to B.A.D. 1 at Burtonwood. The engine test block ran its first radial engine on 6 October 1943 and its last on 7 December 1943, with one hundred and twenty-six of these engines cleared before the changeover. On 20 December the change began to take effect as one hundred and twenty-five enlisted men attended the In-Line Engine School which Allison and Rolls-Royce technical representatives had established at Warton. The course lasted two weeks with numerous films and lectures, while hands-on experience was received in the Engine Repair Section.

It was normal for American bases in England to invite local children to a Christmas party, and for one week between 16 and 22 December 1943 the children visited. Over seven hundred were entertained with base personnel going to great lengths to arrange parties with plenty of food, sweets, chocolate and, of course, chewing gum. The Americans also made sure that every child had a substantial toy. Colonel Moore went out of his way to ensure that things went well and personally greeted every arrival.

Besides the Christmas parties there was a break from routine for some of the other men as an exercise to test the airfield defences was arranged. The 'enemy' was the local Home Guard, and 'by arrangement' all ended well with honour being satisfied on both sides.

By the end of 1943, as seen in the following table, a total of three hundred and thirty-five aircraft were made operational and released to the various units:

Type	Aug.	Sept.	Oct.	Nov.	Dec.	Total	
B-17	3	9	29	82	94	217	
B-24	—	—	—	1	5	6	
P-51	—	—	2	35	62	99	
P-47	—	—	2	1	8	11	
P-38	—	—	1	—	1	2	Total 335

One aircraft accident was recorded in December when on the 11th First Lieutenant G.A. Fencas of the 310th Ferrying Squadron was killed when the P-38 Lightning he was flying, 42–67480, crashed near Longnor in Staffordshire.

CHAPTER 3
1944 – The Big Year

The new year brought a complete reorganisation of working practices at both Warton and Burtonwood to achieve maximum effort from the units, and as the months passed the wisdom of this change was reflected by the massive increase in the production rate. To help achieve the 'production line' methods, B.A.D.A. Headquarters allocated a specific bomber and fighter type to each depot: Warton would be responsible for the P-51 Mustang and B-24 Liberator while Burtonwood specialised in the P-38 Lightning and B-17 Fortress. Both sites would process P-47 Thunderbolts, but B.A.D. 1 would have a greater share of this type. Other types, such as the B-26 and, later, the P-61, would be assigned to Burtonwood, with the Douglas A-20 Havoc and, later, the A-26 Invader accepted at Warton. The C-47 transports were shared between the depots.

As the backlog of aircraft increased it soon became necessary to utilise all the main hangars at Warton. Storage areas were reclaimed. Hangar No. 1 did keep a small storage area, but was used for assembling the smaller aircraft types including L-4s, L-5s, UC-64s and AT-6s; Hangar 2 also housed some of the smaller types and the A-20 Havocs.

In the autumn of 1944 the A-20s gave way to the new A-26 Invaders. Hangar 3 handled B-24s, C-47s and the occasional B-17; Hangar 4 had been mostly for work on B-17s, but with the changeover the B-24s moved in. In addition to overhauls many B-24s were converted for a variety of duties including 'Carpetbagger' operations – the code-name

The unpredictable weather in Britain made instrument training essential even for the most experienced pilot. Here Fay Brandis receives instruction in Warton's Link Trainer.

Besides the normal overhauls and modifications, battle-damaged aircraft had to be repaired as can be seen in this photograph of a B-24 which arrived on 9 February 1944.

Above:
One of the Tiger Moths on loan to the 8th Air Force still wears its R.A.F. roundels although it has already gained a shark's mouth design on the cowling. In the background is the Oxford AS728 which was operated by the 4th Fighter Group at Debden in Essex and, presumably, this visit was in connection with the introduction of the P-51 into that group.

Below:
This B-24, for some obscure reason, had a rear gun turret fitted into the nose position.

Above:
Pulling through the propellers to help circulate the oil in the engine cylinders was a hard but essential job. The task was particularly difficult on the cold winter mornings.

Below:
Orville Wrosch relaxes in the April sunshine while working on a P-47 which had returned to Warton for overhaul and reallocation. The Thunderbolt had served with the 4th Fighter Group until that unit converted to the Mustang.

The World's Greatest Air Depot

A Douglas A-20 Havoc was destroyed by fire during engine runs which were usually carried out near the fire station, just out of the picture. The fire crew were at the other side of the airfield when the aircraft caught fire, and it was completely burnt out by the time they reached the scene.

for clandestine tasks – while others had large fuel tanks fitted in the fuselage to carry much-needed fuel to the Continent. Hangar 4 also coped with the urgent influx of C-47s for engine changes, and a period when P-47s needed wing brackets and the installation of water injection tanks.

Hangar 5 was the P-51 line. The first P-51Bs arriving in England had gone through to have an additional eighty-five-gallon fuel tank installed in the fuselage, and wing brackets were fitted to carry drop tanks. Another task was the fitting of G-suit adapters into the Mustangs. After the Invasion, No. 5 also handled P-47s urgently needed for the fighter-bomber role to support troop movements on the Continent. Hangars 6 and 7 belonged to Flight Test where aircraft were prepared for flight and cleared by the B.A.D. 2 test pilots before being delivered to their operational units. Hangars 31 and 32 were located on the side of the airfield adjacent to the River Ribble. Completed in the autumn of 1944, their main programme did not commence until the following February when they were used for the assembly of Waco CG-4 and CG-15 gliders. These aircraft later took

Right:
Dreamboat *was converted from the B-17E 41–9112 of the 92nd Bomb Group, and this photograph shows powered tail turret*

1944 – The Big Year

*This photograph, taken on 10 March 1944, seems to be nothing out of the ordinary, but to the right of the shot can be seen the experimental B-17 **Dreamboat**. This aircraft had been fitted with a complete change of armament including B-24-type turrets in the nose and tail positions and, although potent, the aircraft was not popular with those who flew it and the project was later abandoned.*

part in the Rhine crossing for the final assault on Germany. Other hangars on the base were not adjacent to the airfield and were used for a variety of work tasks.

Although there were not enough fully trained technicians to staff every hangar, crews were streamlined and spread through the newly acquired facilities. Production progressed so rapidly that the backlog of work soon disappeared, and in April 1944 the teams were becoming so efficient that they were beginning to run out of work. This was due, to a large extent, to the decision of the Eighth Air Force to limit the number of heavy bombers on the airfield to twenty-five. A programme which showed that at least fifteen bombers could be handled every day was forwarded to Bomber Command and the restriction was lifted.

For Flight Test the year started badly with the loss of First Lieutenant Lloyd D. Bingham Jr on 4 January 1944. He was flying a P-51 when the aircraft was seen to roll over and dive into the sea six miles off Blackpool. A search of the area proved fruitless.

An amazing incident happened on 16

Left:
Colonel Paul B. Jackson was appointed Chief of Maintenance on 15 April 1944 and this popular figure is still a prominent member of the B.A.D. 2 Association even at ninety-four years of age.

The World's Greatest Air Depot

Above:
A civilian contractor, Mr Frederick Cooke of Blackpool, was killed on 17 April 1944 when the lorry on which he was riding was struck by a taxying Mustang.

Below:
As some hangars were still used for storage there was no space for arriving aircraft and work had to be done in the open. The construction of new warehouses alleviated the problem.

January when an A-20 aircraft was on engine runs behind the main hangars near the base fire station. During the run, as the power was increased, the aeroplane caught fire, but the fire crew was on exercise on the far side of the airfield, and after racing to the scene they arrived to find the machine completely burnt out.

On 3 March a B-26 Marauder visiting the station belly-landed on the main runway, but as test flying had to continue, parked aircraft had to be moved from the short runways. The Alert Crew was highly trained for such duties and the test flights soon resumed.

On 17 April 1944 a civilian contractor named Frederick Cooke was killed when the lorry on which he was riding was hit by a taxying P-51. The driver and two other men escaped injury, although Mr Cooke, who was riding on the back, was caught by the aircraft's propeller.

The 17th was not a good day for Warton, especially when the top turret guns of a B-24 were accidentally fired. The aircraft was in Hangar 3 for maintenance when the guns went off, spraying the area with fifty-calibre shells. Fortunately nobody was hurt. Later that day, the Engine Test Block caught fire. One day earlier, a P-47 Thunderbolt named *Miss Georgia Peach* woke up most of the station when, just before midnight, an electrician working on the aircraft in Hangar 5 accidentally touched the gun button with all eight guns blasting off part of the hangar roof. The replacement piece can be seen to this day.

After the change to in-line engines, the first Packard Merlin was run in the Engine Test Block on 16 January 1944. The Allison V-1710 engines for the P-38 Lightnings were tested until the following October when the last of 2,586 of that type was run. Despite the heavy work load testing both Merlins and Allisons, the department had an excellent safety record. Apart from the Engine Test Block fire recorded earlier, the most serious blaze occurred on 30 May 1944 in cells No. 5 and 6, trapping three men. Sergeant G.I. Miller received the Soldier's Medal for disregarding his own safety to rescue the most seriously injured man. The only other serious incident happened when a technician received a cut head after being hit by a propeller on 11 June 1944.

On 15 February 1944 Lieutenant-Colonel B.F.W. Heyer became Chief of Maintenance after taking over from the Acting Chief, Major William H. Arnold. Heyer kept that important position until 15 April 1944 when the experienced Lieutenant-Colonel Paul B. Jackson assumed the role, but the former stayed on to assist the new chief until his orders came through for a transfer on 2 May.

The sheet metal workers of the Manufacturing and Repair Department were presented with an important task from orders received on 8 March 1944. Special jigs and tools had to be devised and manufactured to completely rework the main spar of P-51 tailplanes. Master Sergeant Arthur Wade was line chief in charge of the P-51s and, in an excellent period from 9 January to 1 March 1944, his section processed and turned over to the test pilots three hundred Mustangs, all of which were flown without technical failure and were cleared to join combat units.

On 21 March another directive concerning the P-51 arrived at B.A.D. 2. This time it was ordered that all engine bolts had to be removed, given the Rockwell Hardness Test to determine the required tensile strength, and then X-rayed to detect any minute cracks. Any replacement bolts were manufactured on site.

The hard work and pressure to supply aircraft for operations could have taken its toll of the men, so there was a definite need for recreation. The B.A.D. 2 Special Services established three sections including Athletics, Entertainment and Education. Warton had a base theatre with up-to-date films which were changed daily with an audience of 20,000 per month. Music was required and a request for donations to buy instruments received an excellent response, and the B.A.D. 2 Orchestra was formed. The first concert raised over £400 and this was donated to the British Army Navy and Air Force Relief

Above:
Major Joe Stenglein, C.O. of the 310th Ferrying Squadron, shows how things should be done as he flies **Demon Deacon** past the control tower during a mini airshow on 21 March 1944.

Below:
Another view of the **Deacon** during the airshow which had been arranged to show personnel the airfield under a possible air attack.

1944 – The Big Year

A line-up of newly arrived P-51B Mustangs outside the Flight Test hangars on 25 February 1944. The nearest aircraft, 43-6562, became WD-N of the 4th Fighter Group before being lost on 11 July 1944.

A view of the operations room with a ferry crew studying a large-scale map of Britain. Note the recognition models hanging from the ceiling, including a number of Japanese aircraft!

One of the early P-51Bs with Flight Test personnel: (L to R) Jack Kennedy; Spence Thwaites; Pete Orlick in cockpit; Art Kelly; Mac Gattis, kneeling; Bob Gokey; Pee Wee Reese; and Bill Pickering.

The World's Greatest Air Depot

The operations room photographs were taken on 20 April 1944 and this one shows the duty Sergeant at his communications panel. Very pistols can be seen in the wall rack.

The base hospital was gradually extended to over two hundred beds and was recognised as one of the finest U.S. hospitals in Britain. After the war the hospital was used by the R.A.F. as a medical training centre.

Fund. Auditions for singers, dancers and comedians were well attended, and the wide range of talent surprised many.

The building programme which had been progressing through the winter began to bear fruit in March 1944 with the occupation of several new buildings. On the 6th, Warton was visited by Brigadier-General I.W. Ott, Commanding General of B.A.D.A., and during his tour of inspection he ordered the removal of the Internal Supply Section which was still in Hangar 1. This was the last section to vacate the hangars and the move was completed in twelve hours with Modification and Repair taking over, quickly replacing the stock bins and shelves. Supply moved into two new warehouses measuring 1,080 feet by 250 feet which were situated across the Lytham Road.

Another P-38 Lightning was lost on 18 March when Flight Officer William H. Valee of the 310th Ferrying Squadron was fatally injured in a crash near R.A.F. Woodvale.

The strength of the depot continued to increase and by March 1944 totalled 10,408 personnel, but by the middle of the month this figure was reduced to 9,276 with the transfer of men to A.A.F. Station 594 at Stone in Staffordshire. However, the move had been due to a mix-up and most of the men returned to Warton by the end of the month.

During the last days of April it was announced that a War Bond Drive would start with the object being to purchase two P-51 Mustangs. The goal was $114,000, with one of the aircraft being named *Too Bad*, derived from the depot's title. The other aircraft's name was to be selected by an enlisted man after a draw. Amazingly, the drive was oversubscribed by enough to buy a third Mustang. Colonel Moore drew out the

*Photographed in February 1944 the Flight Test officers are: (L to R) Tom Boland, engineering officer; with pilots, Charlie Himes; Burtie Orth; John Willett; Mike Murtha and John Bloemendal. The Mustang in the picture is **Short Fuse Sallee** which was the personal aircraft of ace Richard E. Turner of the 354th Fighter Group.*

With its long nose and restricted vision, the P-51 was a difficult aeroplane to taxi as this photograph shows. Almost half the parked Mustang's wing had disappeared before the unfortunate crew chief was able to stop his charge.

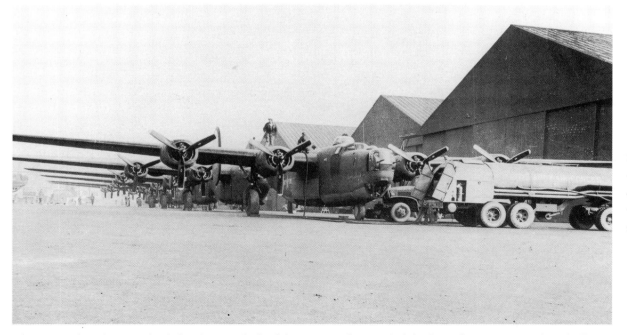

These B-24s have been overhauled and are ready for delivery to combat units of the Second Air Division. The refuelling is taking place behind the main production hangars as ramp space was always at a premium.

The World's Greatest Air Depot

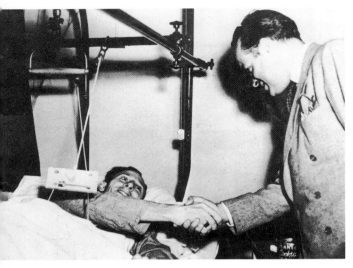

'Salute the Soldier' week in May 1944 proved to be a popular event with a number of parades, displays and exhibitions. Here, crowds gather outside the Warton village hall for speeches by British and American officers and the leader of the parish council.

Above Left:
Technical Sergeant Paul Oberdorf received a great boost when he was visited by Bob Hope while he was in the U.S. Army Hospital at Nottingham. Paul was badly injured by the nose-wheel collapse of a B-26 Marauder, but recovered after seven months to return to Warton in February 1944.

Below:
Fog at Warton forced Jack Knight to divert his B-17 to Squires Gate, Blackpool. Unfortunately, the weather was little better there and the Fortress ended up in the grass. However, there was a happy ending, for as Jack and engineer Spence Thwaites climbed out they realised that just a few yards more and they would have been in a large open cesspool!

1944 – The Big Year

On 30 May 1944 Warton's personnel turned out in force for a ceremony to present three Mustangs, purchased with their war bonds, to the Army Air Force. The three aircraft were named **Pride of the Yanks**, **Mazie R** and **Too Bad**.

names of the enlisted men who would name the second and third aircraft. After discussion with his barrack mates, the first winner, Private Sam Silverman, chose to name the P-51 *Pride of the Yanks*, while Private First Class Stanley Ruggles called his machine *Mazie R* in honour of his mother. The unveiling of the Mustangs and their presentation were made on 30 May at 1630 hours when the majority of the base personnel would be available to watch the ceremony during a change of shifts. Speeches were made by Major-General Hugh Knerr, C.G. A.F.S.C., and Brigadier-General Ike Ott, C.G. B.A.D.A. After the men unveiled the aircraft, Colonel Moore formally presented the P-51s to the combat pilots of three groups on behalf of B.A.D. 2.

On 16 May 1944 Warton received a visit from Lieutenant-General James H. Doolittle. The tour of the base by the famous General was not just the standard visit, but was an excellent public relations exercise as he went out of his way to meet as many of the B.A.D. 2

Left:
*One of the presentation P-51s, **Too Bad**, can be seen on the platform as Major-General Hugh Knerr congratulates B.A.D. 2 on the fine effort of buying enough bonds for three machines, as the target was only two. Seated behind the General are Colonels Paul Jackson and Robert Patterson.*

Above:
A rare photograph of the experimental Mustang X, AM121, which had been fitted with a Rolls-Royce Merlin 61 engine and was identified by the much deeper engine cowling. The aircraft was based at Bovingdon for evaluation by the VIIIth Fighter Command and had been flown to Warton for internal modifications.

Below:
Major Gale E. Schooling, Chief of Flight Test, gives a briefing to one of his sections before the start of a shift. With him are test pilots Jack Knight and John Willett, assisted by Tom Boland, the Maintenance Division's engineering officer. In the background are a P-47 Thunderbolt and a Lockheed Hudson.

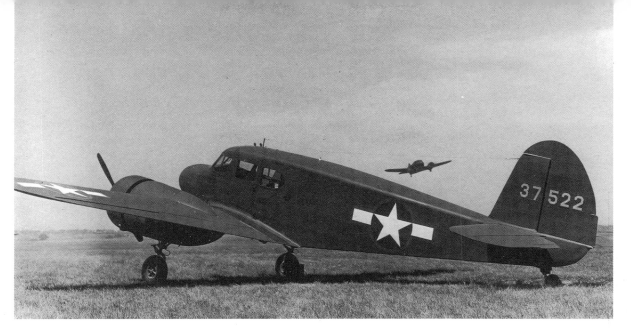

Above:
The Cessna UC-78 Bobcat was a pretty aeroplane and was used throughout the 8th Air Force, but some structural failures were encountered on occasions and the aircraft was not popular with its ground crews. Warton's UC-78, 43–31827, was known as the **Bamboo Bomber**.

Below:
A delegation of Russian Air Force officers visited Warton in May 1944, and here one of the Russians inspects the P-51B 42–106683 which was later transferred to the R.A.F. as SR423, a Mustang III.

personnel as possible. This proved to be an unforgettable experience for those lucky enough to chat with this legendary figure.

Excitement turned to sadness on 27 May when one of Warton's aircraft which had been reported missing was heard to have crashed with no survivors. The aircraft, a Cessna UC-78 flown by First Lieutenant Pliny R. Blodgett, was on a flight to Renfrew, Scotland when it crashed into the hills to the north-west of its destination. There were two passengers on the flight.

By the end of May all the anti-aircraft guns which ringed Warton, plus one which was located on the edge of the ramp opposite Hangar 2, had disappeared, presumably (no one knew at the time) going south to be part of the build-up for the forthcoming invasion.

Major Himes at the controls of the P-51B **Cisco** *over the Irish Sea. The aircraft's code letters show that it belonged to the 356th Squadron of the 354th Fighter Group.*

Chapter 4
The Saga of *Spare Parts*

The most famous aeroplane ever to operate from Warton, at least as far as the B.A.D. 2 personnel were concerned, was a P-51B Mustang by the name of *Spare Parts*. Its story began on a cold, dreary night on 19 February 1944 when the American cargo ship *Spica* tied up in the docks at Liverpool bringing its precious cargo of Mustangs for service in the European Theatre of Operations (E.T.O.).

At first light the following morning work commenced on the unloading of the twelve Mustangs on board the vessel. All the aircraft were P-51B-5-NAs built by North American Aviation at its plant in Inglewood, California, and were complete apart from their tail surfaces, propeller and wingtips which accompanied them in their individual crates. The aircraft were carefully lowered on to specially designed low-loading trailers which

Just one of many! A P-51B Mustang during overhaul and installation of a fuselage fuel tank in Hangar 5 in this photograph dated March 1944.

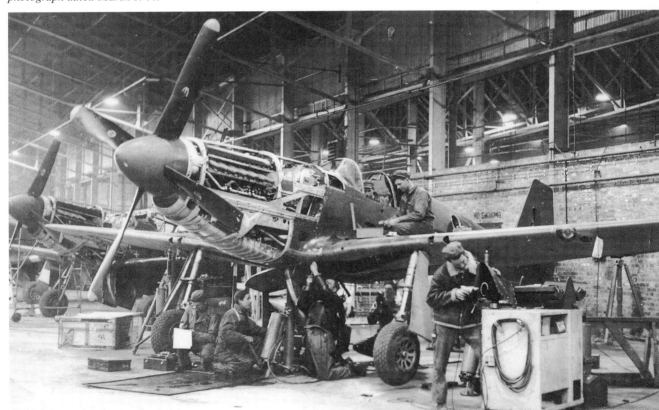

The World's Greatest Air Depot

Warton's own Mustang **Spare Parts** was a great morale-booster for the men of B.A.D. 2, for apart from rebuilding the aeroplane from a written-off machine and then flying on whisky-running missions, the rear seat gave many ground crew members the excitement of flying in a single-engined fighter plane.

The men of Flight Test got together at the end of one shift to have their photograph taken in the May sunshine.

The Saga of 'Spare Parts'

were towed by R.A.F. Bedford tractor units. It was a daily occurrence in the streets of Liverpool to see convoys of trucks towing a variety of American aircraft types from the docks to the airfield at Speke where the Lockheed Overseas Corporation had a facility. The aircraft would be reassembled there and, depending on type, then flown to B.A.D. 1 at Burtonwood or to Warton.

The unloading of the olive-drab P-51s was proceeding smoothly and causing no problems for the experienced dockers and merchant seamen, but as the seventh aircraft was being hoisted from the ship the carrying sling started to slip, tilting the fuselage downwards at a steep angle. Before anyone could rectify the situation, the sling broke completely and the aeroplane crashed to the ground. The Mustang, 43–6623, was hurriedly pulled to one side for a new sling to be fitted to the crane and the unloading to resume as soon as possible. Urgent telephone calls were made to both Speke and Warton with inspectors arriving from both sites to survey the damage. An extensive survey was completed and the final report recommended that the Mustang should be written off. The aircraft would be transported to Warton where it could be stripped down and used for spare parts.

The aircraft arrived at B.A.D. 2 on 1 March 1944 and sat forlornly in the corner of Hangar 5 for a number of weeks without being touched as spares were already coming through plentifully and at a steady rate. Many of the ground crew were aviation engineers and specialists in their own field and, after seeing the Mustang was unwanted, they decided to approach Colonel Moore for permission to rebuild the aircraft in their own time. They were convinced that with an extensive programme the P-51 could be made to fly again, even to enter a combat unit. They were pleasantly surprised when permission was granted, providing that the project was inspected at every stage of the reconstruction. The aircraft was completely stripped down with the badly damaged sections being removed and rebuilt or replaced. As news of the work spread through the base more willing helpers joined the team with the various tradesmen eager to make their own

A general view of Hangar 5 showing accommodation for over twenty Mustangs.

Colonel Moore with the Adjutant's Office personnel. To the Colonel's right is Major F.M. Rothermel, while second from the right is Arthur Loscher, now both members of the B.A.D. 2 Association. Arthur was one of the last at Warton to assist with the closure of B.A.D. 2.

contribution. Inspectors viewed the rebuild phase by phase until the final stage came when the Mustang's Packard-Merlin engine was stripped, cleaned and thoroughly overhauled. Totally satisfied, the inspectors gave the aircraft clearance to fly and Colonel Moore, who was thoroughly impressed by the project, agreed, although he stated that the Mustang must be used as a station 'hack' and must never be delivered to a fighter group.

Naturally the volunteers were disappointed that the aircraft could not be used for the very purpose for which it was intended, but they were excited at the prospect of having their own P-51. Now in natural-metal finish, the P-51 was equipped with all the latest improvements and modifications, but before the first test flight it was decided to remove the oxygen system, relocate the SCR522 command radio, and install a second seat immediately behind the pilot. It seemed only natural to name the aircraft *Spare Parts* as it was estimated that sections of five different aircraft went into the

The busy flight line outside the Flight Test Hangars 6 and 7 with P-47s and P-51s waiting to be tested.

*Ray Dlouhy tops up **Spare Parts** before another 'important mission'.*

rebuild, and the name was duly painted on together with a rather glamorous young lady. The honour of completing the artwork fell to Sergeant Bob Vroman and the finished paint job was admired by all.

The test pilot for the first flight was to be First Lieutenant Jack Knight, or 'Smilin' Jack' as he was known to all after the famous Saturday matinée aviator, and it was he who suggested that one of the rebuild team should accompany him. (Jack would never say whether it was a reward or a threat!) Lots were drawn and radio specialist Sergeant Lee Maudlin was the envy of many as he picked out the winning ticket. The only problem was that the passenger had to enter the aircraft through a window on the left side of *Spare Parts*, but being a slim twenty-year-old Maudlin had no problems. The Sergeant was an experienced flyer on the larger types, but he was the first of Warton's ground crew to fly in a single-engine fighter, and one of his colleagues passed him a brown paper bag 'just in case'.

The Merlin burst into life at the first attempt, and by the time Knight had taxied to the end of the runway the whole of B.A.D. 2 was practically at a standstill as men gathered to watch the take-off. 'Smilin' Jack' certainly did not disappoint them as, after lifting off, he immediately retracted the landing gear and kept the P-51 low for the entire length of the runway before climbing near vertically to 5,000 feet. He began to put *Spare Parts* through her paces, and with Warton's usually busy airspace free of traffic the base personnel marvelled at the display. As Lancashire's July sunshine began to fade, Knight was given permission to land, and after a high-speed pass and a pull-up to slow down, he brought the Mustang in for a perfect landing. During the twenty-five-minute flight the aircraft behaved well and Maudlin did not need to use his paper bag, despite a full aerobatic routine which included nineteen rolls. However, after such enthusiasm it was interesting to note the number of ground crew who decided to pass up the invitation to fly in the P-51 at a later date.

Ed Stoke at the wheel of Air Traffic Control's jeep which was used for patrolling the airfield, with radio contact to the control tower being maintained at all times.

During its service at B.A.D. 2 *Spare Parts* was used for many tasks including VIP transport, delivering urgently needed spares to combat stations, pilot familiarisation, ground crew taxi training, and a variety of other jobs. The aircraft's most important job, as far as the Warton personnel were concerned, was as a 'rum runner', or more accurately a 'whisky runner'. A few times a month a 'training flight' would be arranged to visit Renfew near Glasgow where a nearby distillery would supply bottles of their finest at wholesale prices, and the P-51 usually returned to base with two cases strapped into the rear seat. Extra bottles were even packed into the wing gun compartments as the aircraft carried no armament. Before a flight it was not unusual to see the technicians passing pound notes to the pilot and then, after the aircraft had landed, the same men would be out on the ramp to 'inspect' the Mustang. Some of these flights were followed by a C-47 also on a 'training flight' to Scotland.

Sadly, late in 1944 the aircraft was lost when it crashed after serious engine trouble. Some report that the aircraft went down over the Irish Sea, while others say it was in the south of England; the official records show that the P-51B 43–6623 was 'damaged beyond repair at Liverpool on 20 February 1944'. The other P-51s unloaded that day went on to the famous 4th Fighter Group at Debden, and perhaps without that fateful accident *Spare Parts* could have become famous as the mount of Blakeslee, Gentile or Godfrey or another American ace. However, the Wartoneers are convinced that even in her short life she did more for the war effort than many of her sisters ever did!

Spare Parts *airborne with Joe Bosworth at the controls and one of the ground crew enjoying the ride.*

CHAPTER 5

The Invasion – The Urgency Multiplies

Operation 'Overlord', which took place on 6 June 1944, was the largest combined operation the world had ever witnessed. The air forces had a vital role to play in the landing of the Allied armies on the heavily defended coast of France, but superiority over the Luftwaffe had to be achieved before any invasion could be undertaken. The vast scale of 'Overlord' required continued support, and nowhere more so than at the Base Air Depots.

The news of the landings was announced over Warton's Tannoy system and every man felt he would have to work just that little bit harder. The steady increase in the production rate in the first five months of the year was encouraging, and the result of their efforts was shown in the fact that shortages of aircraft in the 8th and 9th Air Forces had been alleviated and a surplus pool of aircraft were in storage. The need for these aircraft came shortly after D-Day, and in the following months the Aircraft Department was taxed to the utmost. The uncertainty of the English weather was always in the background, but for the pilots of Flight Test it became necessary to test fly over eight hundred aircraft in a month, and some of these needed two or three flights before they could be cleared for release to combat units. Because of some days lost through the weather as many as fifty aircraft had to be

This B-24, having been completely overhauled, is now being prepared to be towed out on to the flight line. After a test flight the aircraft will be made available for delivery to an operational station.

53

The World's Greatest Air Depot

The visiting B-26 Maurauder **Lonesome Pole-Cat**, *which had completed one hundred missions, created a great deal of interest among the personnel of B.A.D. 2.*

The Armament Department was one of B.A.D. 2's many specialised units and this shot shows the gun turret repair section. During the middle of 1944 the shortage of turrets became critical with battle-damaged aircraft having holed or smashed turrets. Many were rebuilt or recovered from the salvage section to be overhauled and put back into use.

tested on a single day, and considering that only ten pilots were assigned to the section it was a magnificent achievement.

One of the most difficult tasks during that period fell to the Alert Crew who had the responsibility of servicing and dispersing aircraft around the airfield and maintaining all completed machines in a 'ready' state, which meant daily and pre-flight inspections. With only a small number of aircraft it would have been an easy task, but sometimes, with more than eight hundred aircraft and three hundred of those ready for delivery, it called for the highest degree of competency. Great skill and care was needed in the handling of aircraft to avoid damage to others in the tightly packed parking areas.

In conjunction with all this activity Air Traffic Control and Warton tower, call-sign 'Farum', accomplished an outstanding feat just to keep the traffic moving in the midst of what could easily have been chaos. Besides the test flights there were aircraft arriving to be processed, others departing after their clearance, and many visitors on transport flights, liaison work or, believe it or not, arrivals on cross-country training flights. Because of the urgency, aircraft would be given clearance to take off even though other aircraft were on the approach, and test flights had to be kept to a minimum although with the same thorough examination. An illustration of this

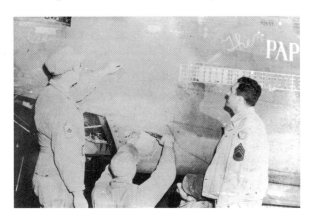

Master Sergeant George 'Killer' Kane and Technical Sergeant John Blainblatt check the standard of Sergeant Orrie's sheet metal work. The aircraft, under repair in Hangar 4, was **The Paper Doll** *of the 93rd Bomb Group stationed at Hardwick in Norfolk.*

was when First Lieutenant 'Pete' Swank was lined up for take-off in a P-51 while the aircraft ahead, a B-24 flown by First Lieutenant Mike Murtha, had climbed to about two hundred feet and was making its first turn. Murtha called the tower asking for landing clearance and, thinking he was in trouble, Swank called on his radio to see if Murtha had a problem. Back came the reply, 'Nope, it got off the ground, the wheels came up and time's a-wastin'.' On many occasions

The Invasion – The Urgency Multiplies

Lieutenants Guinn and Threadwell by a British-built Miles Master which was used by the U.S.A.A.F. as a fast communications aircraft.

The Deacon *buzzing the field after returning from ferrying collection. As aircraft were delivered to various bases the Marauder would make a round trip to pick up the delivery pilots.*

crew chiefs flew as co-pilots on B-17s and B-24s to enable as many aircraft as possible to be checked out.

In the month of June, because of the intensity of the flying, the possibility of an accident loomed large, but it was the unexpected which made it a bad month. On 12 June 1944 a crew chief was taxying a newly processed B-17 across the main runway when he thought he saw another aircraft about to land. He jammed on the brakes forcing the aircraft to stand on its nose, which completely crumpled it back to the cockpit windscreen. There was no aircraft landing and it was assumed that the engineer had seen a bird out of the corner of his eye, but whatever the reason the Fortress would be staying at Warton for a much longer time than expected.

The first of two fatal accidents during the month also happened on the 12th as Second Lieutenant Bill Clearwater was climbing away from Warton in the P-51D 44–13405. It was to be a routine test flight, but after twenty minutes the aircraft passed over the airfield at 3,000 feet and observers heard a distinct change in engine note as power was suddenly applied; looking up they saw the aircraft diving vertically with the starboard wing, which had detached from the Mustang, fluttering to earth like a falling leaf. The aircraft buried itself into the tidal flats of the River Ribble and Major Himes and Captain Tom Boland drove up to the Lytham lifeboat

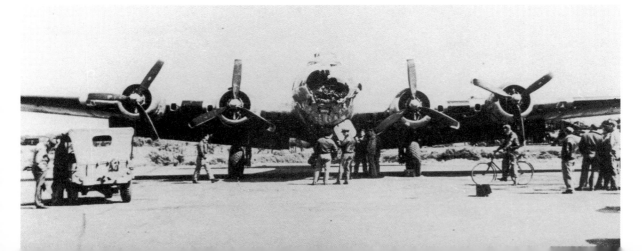

On 12 June 1944 this B-17G nosed over as it taxied across the runway when the crew chief at the controls slammed on the brakes as he thought another aircraft was approaching. His colleagues teased him that it must have been a seagull, but whatever it was the Fortress stayed at Warton for much longer than expected.

The World's Greatest Air Depot

Above:
Chief Test Pilot Charlie Himes pilots this P-47D which had returned for overhaul. The aircraft, wearing the markings of the 352nd Fighter Group, was the mount of Bill Whisner, one of that unit's top aces.

Below:
Bill Clearwater test flying one of the new P-51Ds. Sadly, he was to lose his life in the crash of 44–13405 just days later.

The B.A.D. 2 test pilots practised their formation flying techniques whenever possible, and here four P-51Ds formate on a B-24 from which the photograph was taken.

The same formation returning to Warton flies low over the streets of Lytham. They would then change position to echelon, fly over the airfield, and peel up to decrease speed before getting into the circuit for a stream landing.

The World's Greatest Air Depot

This shiny new P-51D was delivered to the 356th Fighter Group at Martlesham Heath in Suffolk, to become PI-Z of the 360th Squadron.

The beautiful lines of the De Havilland Mosquito XVI, NS635, on 17 June 1944 with Charlie Himes in command and John Bloemendal as co-pilot.

Unfortunately an electrical failure brought an end to the flight and the aeroplane's career, luckily without injuries to either crew member.

Jack Knight was more than happy to accommodate two pretty members of the American Red Cross when they requested to be shown the new Mustang. The small stencilling on the fuselage shows the lucky aircraft to be 44–14429. However, it was not all play as Jack flew 1,166 times in his two years at Warton.

'Smilin' Jack' Knight in a typical pose. Jack maintained that he had two loves in his life, 'airplanes and ladies, in that order!' For information, the Liberator belonged to the 486th Bomb Group.

This new P-51B awaiting a test flight was later delivered to the 339th Fighter Group at Fowlmere, Cambridgeshire, becoming 6N-T of the 505th Squadron.

house where the boatmen were unaware of the tragedy. The two officers went on the boat to the scene of the accident where the remains of Bill Clearwater were recovered. The lifeboat logbook records the call-out as at 2100 hours. Most of the twisted wreckage was recovered and laid out in one corner of Hangar 6 and the structural failure point was soon discovered, but the cause remained a mystery.

The following morning test pilots Jack Knight and John Bloemendal took off in P-51Ds and climbed to altitude where each ran a critical eye over the other's aircraft to check for any tell-tale signs of a failure. Even though the doomed aircraft had been flying straight and level at the time of the disaster, both pilots gave the Mustangs a thorough work-out with rolls, loops, dives and pull-outs, stressing the aircraft more than on a normal test flight. As the pair taxied up to Hangar 6 they must have drawn the conclusion that the accident was a one-off as there were no problems with either of the tested P-51s.

Four days later, on the 17th, a ceremony was held in the Post Theatre as Brigadier-General Isaac Ott C.G. B.A.D.A. presented the Legion of Merit to Lieutenant-Colonel William H. Arnold for outstanding service as Acting Chief of the Maintenance Division until the appointment of Lieutenant-Colonel Heyer. After General Ott had departed Major Himes and First Lieutenant John Bloemendal were scheduled to take the newly arrived De Havilland Mosquito for a routine test flight. The sleek Mosquito PRXVI, NS635, was the type being supplied by the R.A.F. to the 8th Air Force and B.A.D. 2 were to familiarise themselves with the aircraft. Himes and Bloemendal got to 3,000 feet and feathered the left propeller to check the Mossie's single-engine performance. The aircraft was flying slower than expected and Major Himes applied more power to the right engine to increase the speed. As he did so the aircraft started to roll to the left, and as he tried to unfeather the left prop it would not move. He put on full right aileron and right rudder and

The Invasion – The Urgency Multiplies

Above:
Burtie Orth at the controls of a Piper L-4 Grasshopper on a routine flight. This type would become very familiar at Warton as the depot assembled two hundred and forty-three of the small liaison aircraft. He was soon to lose his life in the second fatal P-51 accident in June 1944.

Below:
The massive aircraft fuel servicing tankers of the U.S.A.A.F. carried 4,000 U.S. gallons. This vehicle was in Dispersal Area 2 to refuel a B-24 which was awaiting delivery to an operational unit. Also in the photograph are a Douglas A-20 and a B-17 transport aircraft with Hangars 1 to 7 in the distance.

Above:
The newly arrived P-51Ds required a number of modifications before delivery, including a dorsal fillet to the fin to improve longitudinal stability as the rear fuselage had less area after the introduction of the clear-view canopy.

Below:
After the tragic crashes of the P-51Ds of Bill Clearwater and Burtie Orth, it was left to the engineers of B.A.D. 2 to try to find the cause. They thoroughly tested one of the breed and solved the mystery with their conclusions, no doubt, saving the lives of P-51D pilots in the future.

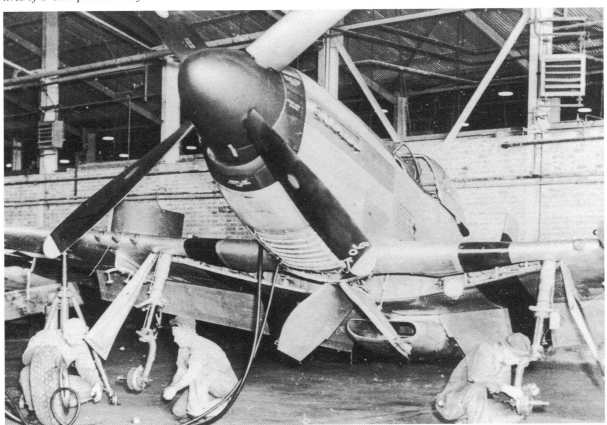

The Invasion – The Urgency Multiplies

only then did he start to get the aeroplane under control, but only just. The Mosquito was then steered towards Warton, but with still more right power the aircraft wanted to continue the roll. Himes lined up with the main runway and lowered the undercarriage, but as he did so the aircraft almost stopped. He quickly raised the wheels to pick-up speed, but the Mosquito bellied-in almost immediately. As the chief test pilot, Charlie Himes was naturally embarrassed about the whole episode and an investigation showed that a flat battery caused the problem with the unfeathering.

An amusing event connected with this incident happened later that evening when Lieutenant-Colonel Jackson announced in the Officers' Club that henceforth Major Himes would have the nickname 'Flit'. For a moment the name was lost on the assembled officers until it was pointed out that 'Flit' was the name of a spray for killing mosquitos! To this day Charlie Himes is known as 'Flit' by members of the B.A.D. 2 Association.

Just ten days after the Mosquito crash, disaster struck again. Second Lieutenant Burtie Orth was due to test fly a P-47 Thunderbolt, but he was advised that the engine power was down on performance and so he decided to test one of the P-51Ds ready to fly. The aircraft took off and climbed to 7,000 feet into a beautiful clear sky when the right wing detached and started falling to the ground. The incident was witnessed by personnel from Site Eight at the base who saw the aircraft flying straight and level as the wing broke off, but it was different from the earlier crash in that the fuselage seemed to take longer to hit the ground as power was applied at intervals. The general feeling was that Orth could have baled out, but he delayed his exit too long as he tried to prevent the Mustang hitting some houses. Local police and National Fire Service (N.F.S.) members located the pilot's

*Two P-51Bs await clearance for take-off for a delivery flight to the 361st Fighter Group at Bottisham in Cambridgeshire. The rear aircraft, 42–106835, became **Bald Eagle**, coded B7-E, of the 374th Squadron while the other, 42–106830, joined the 375th Squadron as E2-H.*

The B-24H 42–7521 **Poop Deck Pappy** *served with the 392nd Bomb Group at Wendling in Norfolk.*

body and it was Major Himes who had to drive to Blackpool to inform Orth's wife of his death. He had married local girl Freda Jones only three months earlier, and she had just told him she was to have a baby in the following spring.

The wreckage of Orth's Mustang was taken to Hangar 6 where an inspection found the same evidence as Clearwater's machine, but once again there were no obvious clues as to the precise nature of the structural failure. The inspectors of Hangar 5 were given the task of finding the answer to these accidents, and a new P-51D was placed on wing jacks with an undercarriage retraction test carried out to see if any malfunction could be duplicated. Hangar 5's P-51 experts – Buster Beadle, Ted Tryba, Bill Niemi, Al Canel and Bob Porche – were detailed to study these tests. It was found that the brake lines on both the crashed Mustangs had somehow bent around the undercarriage shock absorbers; the landing gear had dropped during the flight and had twisted completely around.

Here **Poop Deck Pappy**, *with slightly modified nose art, is seen in Warton's Hangar 4 awaiting 'the treatment'.*

The Invasion – The Urgency Multiplies

The Canadian-built UC-64 Norseman, originally designed for bush flying, proved to be a fine workhorse for the U.S.A.A.F.

The Nissen huts of Site 10, left, and Site 11, right centre, are clearly visible in this photograph by Jack Knight. The village of Freckleton is at the top right of the picture.

*One of the 310th ferry pilots receives the paperwork for a P-51D before delivery. For some reason **The Comet** had been stencilled on to this new aircraft. It was unusual for an aircraft to be named before delivery.*

The World's Greatest Air Depot

The Packard Merlin assembly line seen in June 1944. The engine specialists had a fine record with very few failures during the test runs.

The landing runway was 08 on 30 July 1944 when this Piper L-4H was parked near the control tower.

With the test running, the pilot's checklist was implemented as though the aircraft was in flight, and with nothing showing after two hours it was decided to take a coffee break.

With the gear up and the undercarriage lever in the neutral position, but with the hydraulic system in a cruising flight condition, the inspectors started to leave for their break. As they did so one of them noticed that one of the wheel doors had dropped down slightly with the wheel resting on it. In flight the airflow would have been enough to rip this

Warton's technical area from the air early in June 1944 shows the vast assortment of aircraft either awaiting processing or ready for delivery. Note the black 'Carpetbagger' B-24s.

On a busy airfield taxying accidents are always possible: this visiting Miles Master II, AZ600, ran into a stationary jeep. There were no injuries, apart from the pilot's pride.

The Invasion – The Urgency Multiplies

open for the undercarriage leg to extend. A number of changes had been made to the P-51D model of the Mustang including the addition of an extra machine-gun in each wing plus a number of other features with the wing being more highly stressed than the earlier P-51B and C models. Presumably as a weight-saving measure the undercarriage up-locks on the earlier Mustangs had been omitted from the D model and this allowed the wheels to bounce about in their wells with only hydraulic pressure to hold them when the undercarriage was selected up. Any surge or fall in pressure would allow them to force the wheel doors open in flight, extending the wheels and, at a reasonable speed, ripping the wing off. The bouncing around would, no doubt, have caused fatigue and the resulting structural failure. The Warton findings were quickly transmitted to the manufacturer and to U.S.A.A.F. Headquarters with the result that P-51B type uplocks were retrofitted to all D models and the modification incorporated on all new aircraft being built at North American Aviation. The expertise of the Warton engineers had unravelled the mystery, but the price paid in advance was the loss of two young test pilots. If the problem had been allowed to continue unchecked it would, no doubt, have cost the lives of many more aviators.

With the large volume of air traffic always in evidence at Warton the controllers could have been excused for missing the departure of a UC-64 Norseman on 30 June 1944. One controller thought that the other had handled it and vice versa, but in truth the flight was unauthorised and this did not come to light until B.A.D. 2 received a telephone call from the airfield at R.A.F. Valley in Anglesey, North Wales, to report that the aircraft had landed there. The Norseman had flown low over the control tower and touched down on

*Unknown to the men working on **Male Call** in Hangar 4, this B-24 was flown many times by the film actor James Stewart while he was serving with the 445th Bomb Group at Tibenham in Norfolk.*

The World's Greatest Air Depot

the grass before crossing an unused runway to stop in the grass on the far side of the airfield. It was discovered that the 'pilot' was an American private who had taken the aircraft from Warton to fly it to France where his brother was fighting with the Allied ground forces! Needless to say, he was arrested and returned to Warton where he was court-martialled. He was then sent to an American army prison where he served his sentence, only for the war in Europe to have ended before he was released.

Another unusual incident happened on the following day as First Lieutenant G.E. Bachelor of the 310th Ferry Squadron was flown down to a 9th Air Force base on the south coast to bring to Warton a P-47 Thunderbolt for instrument repairs. As the aircraft did not appear it was posted missing, and checks with other bases proved fruitless. It was an order that if a pilot was missing for twenty days his documents must be forwarded to headquarters in London. On the morning of the twentieth day the 310th received a telephone call from Bachelor to report that he was safe and well and had just landed back in England. As the story unfolded, it was found that he had taken off and flown directly into bad weather, and as the P-47's instruments were faulty he had gone in the wrong direction. Still in the murk, he had continued until his fuel ran out and had taken to his parachute, coming down in a part of France still under German occupation. Luckily he was rescued by the Resistance who eventually returned him to the Allied lines. He later admitted that as a ferry pilot he had certainly never expected such excitement.

It was a great credit to all concerned that the Warton air traffic did keep moving as it did because at the end of May the 829th Engineer Aviation Battalion arrived at the

Visits by entertainers and sporting personalities were always welcome, and the arrival of world boxing champion Joe Louis and title contender Billy Conn on 6 July 1944 was particularly well received.

The boxers pose with Colonel John Moore, Commanding Officer of B.A.D. 2. Billy Conn returned in September to spar with base personnel.

The Palace Hotel at Southport was a rest centre for combat crews of the U.S.A.A.F. with accommodation for fifty officers and one hundred enlisted men. Occasionally Warton personnel would stay while visiting friends from other 8th Air Force units. Note the air raid shelters in the hotel grounds.

The Invasion – The Urgency Multiplies

*Steam cleaning the number one engine of the 446th Bomb Group's Liberator **The Spirit of 77**.*

Nothing was ever wasted either from scrapped aircraft or worn-out items as the row upon row of shelves in the vast salvage warehouse will testify. Many recovered items were repaired and made usable once more.

The Aircraft Salvage Department did a magnificant job saving just about everything they could from scrapped aircraft. This A-20 looks worse for wear, although the team had only just started.

The construction of the urgently required warehouses for the Supply Department was almost complete when this photograph was taken on 18 July 1944.

base to renew the perimeter track which had badly deteriorated through the sheer volume of aircraft taxying over it. The original design of the airfield was as an R.A.F. fighter station and the taxi track had remained the same even though the runways had been lengthened and strengthened. The weight of numerous B-17s and B-24s trundling over it had made the track progressively worse.

Besides the Invasion, one of the highlights of the month was the appearance on the twenty-eighth of the new P-61 Black Widow night-fighter. The aircraft landed at Warton only two days after its existence had been officially announced by the U.S. War Department. Naturally, there was considerable interest in the aircraft not only among the flying personnel but the air-minded ground crews too.

At 0600 hours on 1 July 1944, Private Avery A. Abraham had the dubious distinction of being the first soldier ever to go A.W.O.L. from Warton. He returned of his own accord on the 4th, but was brought to trial, convicted and received a five-month confinement sentence to be spent in the station guard room.

Above:
This shot of the B-24J **Lucky Leven**, 44–48820, of the 467th Bomb Group well illustrates some of the modifications completed at B.A.D. 2, including side armour for the pilots and extra blister windows for the navigator.

Below:
The long shifts and pressure of work after D-Day took their toll and short breaks had to be taken where possible. The photograph shows Ray Dlouhy grabbing a few minutes' rest in the corn stooks while working on the B-24J 42–51435.

Above:
Peter Manaserro helps Jack Knight with his parachute before another test flight. During July, test pilots flew well over one hundred times each in a variety of types as the pressure mounted for more combat aircraft after the invasion.

This B-24H 42–50682 was modified by B.A.D. 2 for the 'Carpetbagger' role and served with the 801st Bomb Group, a special operations unit based at Harrington in Northamptonshire.

Test pilots Pete Swank with Bill Watters in the cockpit, compare notes on a P-51D before the latter pilot launched into the Wild Blue Yonder.

This small group is some of the sheet metal men of Hangar 4. Standing in the centre is George Gosney, a prominent member and past president of the B.A.D. 2 Association.

After the earlier departure of the station's anti-aircraft guns, the 494th AAA Gun Battalion (Mobile) also left Warton for service on the Continent. A number felt that this left the airfield exposed to air attack, but as the Luftwaffe's raids on England had dwindled to a handful it was unlikely that B.A.D. 2 would ever be attacked. It was always strange that the Germans never bombed Warton or Burtonwood as one large attack on each could have caused untold complications for the push into occupied Europe, especially with so many aircraft on both the airfields at one time.

Morale-boosting visits by entertainers or sporting personalities were always enjoyable, and on 6 July the famous boxers Joe Louis and Billy Conn called in at Warton. The pair had lunch with men of the 829th Engineers before speaking to a gathering in front of the instrument shop. Before their departure they promised to return to give an exhibition of boxing, and this news was received with great anticipation by the Wartoneers.

Award ceremonies occurred at regular intervals, but one which gave much pleasure to Colonel Moore and the Warton personnel was the presentation of the Soldier's Medal to Sergeant Terrence J. Miller. Sergeant Miller was returning from an evening in Preston when he spotted a fire raging in a private house. Without regard to his own safety he bypassed onlookers, forced his way into the house and up the blazing staircase, leading three women and three children to safety. 'The prompt action and courage displayed by Sergeant Miller was in keeping with the highest traditions of the United States Army and of Base Air Depot No. 2', read the citation.

As the aviation engineers were constructing the new perimeter track, parking spaces for aircraft were at a premium, and on 14 July a B-24 was running-up for a pre-take-off check when a visiting R.A.F. Tiger Moth attempted to taxi through a tight space behind the Liberator. The Tiger was directly behind the bomber as the American pilot gunned the engines, and the blast completely lifted the little biplane up into the air and flipped it over on to its back. B.A.D. 2 personnel rushed to help the British pilot, but he was still strapped in the cockpit and asked if the Americans could right the aeroplane as the fuel was running out and he felt 'rather uncomfortable'.

The Aircraft Salvage Department usually had a straightforward job taking damaged aircraft apart to preserve the good spares, and it was undoubtedly a rewarding task as much taxpayers' money was saved by this system. However, it could occasionally be dangerous. One day, relieving a B-24 of its top turret, a large bullet hole was found, but with no exit hole, and upon investigation it was found that a high-explosive cannon shell was lodged in the turret frame with the detonator still intact. One C-47 arrived and was to be scrapped, but in the rear fuselage was a German propeller box which, when opened, contained German land mines and about twenty rifles. The Ordnance Department was quickly called to the aircraft to collect the box. Once disabled, the rifles would have made excellent souvenirs, but they were never seen again.

The urgent need for P-51 Mustangs was always pressing, but at no time more so than in July 1944 when literally hundreds were on the airfield. The ramp near the control tower had two lines of Mustangs running the entire length with others dispersed around the base awaiting their turn to go through Hangar 5. As soon as one P-51 was pushed out of the north door another would be moved in at the south end. The tower ramp was also used for visiting aircraft and one such Mustang arrived one afternoon, but the weather clamped down forcing an overnight stay. The young fighter pilot, not wanting to waste the opportunity of having a night out in Blackpool, set out for the town to soak up some British culture. Meanwhile, back at Warton his Mustang somehow got mixed up with the others and was pushed into Hangar 5 with the technicians taking off everything that was removable. The cowling, canopy and just about everything which could be unscrewed or unbolted was stripped from the aeroplane.

The following morning the pilot went in search of his machine and was horrified to see that it was not where he had left it. In colourful language he demanded to know what had happened to his P-51, but he was calmed down by the hangar chief who took him for a coffee while he explained what the problem was. Except for the serial numbers all Mustangs looked alike and it was an easy mistake for the mechanics to make. When they 'found' the aircraft it was in the middle of receiving all of the latest combat modifications and all hands turned to getting the Mustang completed in a shorter time with the mechanics, armourers and electricians swarming all over it. Before the end of the afternoon the P-51 was in its original parking spot, but once again the weather turned against them so it was off to Blackpool in quick time for the pilot. The day after he really was on his way, but before he closed the canopy he was passed a note which read, 'Dear Sir, Due to a slight error on our part your aircraft has been completely overhauled, modified and updated. It is now better than the day it left the factory. All of the necessary information has been recorded on the aircraft's record sheet. We hope any inconvenience did not make you feel less happy with our hospitality.' The pilot quickly scribbled a note saying, 'Thanks fellas. Had a great time in Blackpool . . . Your service is good too! This must be the greatest place in the Army Air Force. Signed, Joe.'

Word was received at Warton that two of

The Invasion – The Urgency Multiplies

Here's another one! **Winged Fury** *is towed out of Hangar 4 by Sergeant Howard Tice, and in true military tradition the aircraft's paperwork is being passed over from Technical Sergeant Keller to Captain Ora Bard.*

One of the greatest morale boosters to any base was the visit of Glenn Miller and his orchestra with the great band arriving at Warton on 13 August 1944. They played for the troops on the following day.

Ray McKinley thrilled the audience with some excellent drum solos and his special arrangements. The evening before he had joined in with B.A.D. 2's band 'The Yankee Clippers', who were playing in Blackpool.

the P-51s bought during the Bond Drive in May had destroyed enemy aircraft: the *Mazie R*, named by Stanley Ruggles, had shot down an Me 109 while serving with the 357th Fighter Group, and *Pride of the Yanks* had destroyed two enemy fighters while on an escort mission to Leipzig.

On 13 August 1944 two C-47s arrived at Warton carrying Captain Glenn Miller and his famous orchestra. The musicians stayed the night at the base, Miller sharing a room with Major Fred Jacob who had the privilege of informing him that word had just been received that the band leader was now Major Miller. While Miller was catching up on his sleep, other members of his group including Ray McKinley went off to Blackpool where they joined the B.A.D. 2 band 'Yankee Clippers' who were playing in the resort's Spanish Hall. The following day the Glenn Miller band played for over 10,000 of the base personnel from a platform in front of Hangar 4. A wonderful programme was presented, but an untuned piano supplied by Warton's Special Services was found to be unusable. Technical Sergeant Ray McKinley was the highlight of the show with his drum playing and other special arrangements, with the

Left:
Every inch of space and vantage point was taken for the Miller concert which was staged outside Hangar 4. There were even men on the roof of the next hangar!

concert lasting from 1600 to 1705 hours. The band again stayed the night before flying off in two B-24s bound for B.A.D. 1 at Burtonwood.

During the month a policy was put into effect designed to bring the Warton personnel closer to the war. Individuals were picked from all units to form parties of twelve to fifteen who were then flown to an operational base in southern England where they could become acquainted with the routine on such a station. The chance of getting first-hand information on the value of their work proved to be excellent in raising *esprit de corps* and the efficiency not only of those selected to go, but of eager listeners who awaited the stories of what they had seen.

This excellent view of Warton's busy ramp, taken from a Piper L-4, shows the newly arrived A-26 Invaders outside the main production hangars.

CHAPTER 6

The Freckleton Disaster

The morning of 23 August 1944 was just like any other at Warton as the aircraft movements, although lighter than usual, were continuing at a steady pace. The weather was fine with some broken cloud but with the prospect of rain showers later in the morning, and as this type of weather was not unusual on the Lancashire coast there was no real concern among the air crews. Just after 1030 hours 'Farum', Warton's control tower, received a priority radio call from Burtonwood advising that a violent and fast-moving storm had passed B.A.D. 1 and was heading in the direction of Warton. It was recommended that any aircraft flying in the area should return to the airfield and all hangar doors should be closed.

The two B-24s on test flights from Warton at that time were radioed and advised to land as early as possible, but as the first aircraft flown by First Lieutenant John Bloemendal made an approach to Runway 08 with the visibility was dropping by the second. Bloemendal advised that he would abort the landing and go around, but at approximately 1045 hours the Liberator crashed into Holy Trinity School in Freckleton, slewing across a main road and demolishing a small snack bar

Just nine days after the Miller visit Warton was in the news again, but for a much different reason. A tremendous storm forced the Liberator 42–50291 to crash into the infants' section of the Holy Trinity School in the village of Freckleton causing great loss of life. The photograph shows the aircraft's fin bearing the B-24's serial number.

One of the Liberator's main wheels lies among the still blazing wreckage.

known as 'The Sad Sack' on the opposite side of the road. The aircraft had flown into a violent downdraught which forced it down into the school's infants department. A reliable witness reported that the bomber had been struck by lightning which broke the aircraft in two before it plunged into the school. The aircraft immediately burst into flames with the school and the surrounding area engulfed by a sea of fire.

The sound of the crash brought a rush of people, soldiers and police to the scene and Warton's fire department was at the crash within minutes. Local stations of the N.F.S. responded rapidly and rescue work commenced immediately, and as children were freed from the debris they were taken to Warton's hospital for treatment. Sadly, a great many more were beyond treatment and it was the painful duty of the rescuers to remove the tiny bodies from the devastation which once was their school. They were taken to a temporary mortuary for identification by their bereaved. The death toll eventually reached sixty-one when little Maureen Clark died on 4 September: they comprised thirty-eight children, nine British civilians including two teachers, four R.A.F. sergeants, and ten U.S.A.A.F. personnel, which included the three crew members of the Liberator.

The fateful Liberator's pilot John Bloemendal, seen here in the cockpit of a P-51B, was an experienced aviator and an excellent test pilot. He died along with crew members Kinney and Parr. With him in this photograph are Charlie Himes and line chief Art Wade.

The funeral of the majority of the children and one of their teachers, Miss Jenny Hall, took place on the 26th with the burial taking place in the churchyard of Holy Trinity Church just a short distance from the site of the accident. The mass grave was dug by Warton personnel and there were over five hundred floral tributes from the still stunned villagers and the soldiers of B.A.D. 2. General H.H. Arnold, Commanding General of the

The Freckleton Disaster

The Sad Sack café was a popular haunt for Warton's personnel, but it was completely destroyed as the bomber slewed across the main Lytham Road after hitting the school. Besides the cafe's owners both U.S.A.A.F. and R.A.F. men were killed.

Quoted as 'The saddest day of their lives', Warton's troops carry the children's tiny coffins to their last resting place in the grounds of Freckleton's Holy Trinity Church.

U.S.A.A.F., cabled from Washington D.C. that he desired to be represented at the funeral, and Brigadier-General Isaac W. Ott C.G., B.A.D.A. took his place.

The investigation board made the following report:

The cause of this accident is unknown. It is the opinion of the Accident Investigating Committee that the crash resulted from the pilot's error of judgement of the violence of the storm. The extent of the thunderhead was not great and he could have flown in perfect safety to the north and east of the field. When the approach to land was made the pilot saw that conditions were too bad and he attempted to withdraw, but the violent winds and downdraughts must have forced the airplane into the ground before he could gain sufficient speed and altitude. The statement by First Lieutenant Manaserro who was flying at the same time and place describes the difficulties he experienced in flying back out of danger. It is possible that in the rough air structural failure occurred, however, no conclusions could be drawn from the examination of the wreckage as the airplane was so completely demolished.

RECOMMENDATIONS

That all pilots who are gaining most of their flying experience in England (subsequent to flying school) be emphatically warned about entering thunderstorms or flying under thunderheads. The dangers of such practices are easily learned in the southern United States where intense thunderstorms are frequent, but in the British Isles a pilot is led to believe that thunderstorms are mere showers, hence the rare, severe type is the most apt to trap him.

The statement by First Lieutenant Peter Manaserro referred to in the report of the board was as follows:

At approximately 1030 hours Lieutenant Bloemendal took off on a local test hop in B-24 No. 50291 [B-24H 42–50291, ex-490th Bomb Group]. Immediately following Bloemendal, I took off in B-24 No. 1353 [B-24J 42–51353, ex-93rd Bomb Group]. Weather conditions at the time were favourable, vis. of several miles and a broken ceiling of over 1,500 feet. There appeared to be scattered showers in the area with occasional lightnings. After take-off we headed north and were flying at approximately 1,500 feet. Lieutenant Bloemendal and myself were both standing by on V.H.F. when Bloemendal called my attention to the cloud formation towards the S.S.E. It was a very impressive sight and looked like a thunderhead. Shortly after this, 'Farum' called Bloemendal and ordered him to land immediately. He then asked the cause and was informed of approaching weather. We turned towards base and I was flying off Lieutenant Bloemendal's right wing about one hundred yards out. As we drew near the field I moved further out to be in position to land number two. We let down to five hundred feet and about four miles N.W. of the field we encountered rain and it became heavier with less visibility as we neared the approach to Runway 08. On the base leg Bloemendal let down his gear and I did the same. Shortly after this I lost sight of his aircraft. I was watching the ground directly below me as there was no vis. ahead. As I flew over Lytham, I started a left turn to start the approach. At this time I heard Lieutenant Bloemendal notify 'Farum' that he was pulling up his wheels and going around. I was then over the wash and could not see the ground and had to fly on instruments. I then called Bloemendal and told him we had better head north to get out of the storm. He answered 'OK'. I then told him I would take a heading on 330°. He said, 'Roger'.

That was the last I heard from Lieutenant Bloemendal. I flew about four or five minutes on a heading of 330° before breaking out of the storm. I then called Lieutenant Bloemendal and asked if he was OK, but did not get a reply. I contacted 'Farum' and told them to call him. They were also unable to contact him. In my opinion the storm was too severe even for a most experienced pilot. I did not have complete control of my aircraft while in the storm. I was fortunate to find ground with my aircraft in almost a level attitude and no obstructions.

The storm referred to in the statement of Lieutenant Manaserro was an unusually severe one which travelled north from South

Peter Manaserro chats with a fellow pilot during an 'essential training flight' to Villacoublay near Paris. Needless to say, supplies of perfume, cognac and silk stockings improved at Warton!

Wales . . . past Lancaster in north Lancashire and subjected the districts through which it passed to much devastation with crops beaten down, walls knocked over and roofs ripped off. It was surmised that the aircraft might have been struck by lightning as reported, but the investigating team could find no real proof. Any crash is regrettable, but this proved to be a disaster as it was the worst place in the village of Freckleton for the aircraft to come down, and the storm which caused the crash had cleared the area in less than half an hour.

Even today the crash ranks high on the list of accidents involving loss of life on the ground and, Freckleton itself, after losing thirty-eight children, has a large age gap in one section of the local inhabitants. First Lieutenant John A. Bloemendal was a pilot of great experience who had arrived at Warton as a qualified test pilot. First Lieutenant Peter

The Freckleton Disaster

Warton's base hospital did outstanding work in aiding the survivors of the bomber crash and both doctors and nurses deserved and received the highest praise. Pictured are (L to R) Nurse 'Skip' Romanowski, Dr Sharpley and Chief Nurse Nell Russell.

Manaserro was also fully qualified and had arrived at Warton in April 1944 after flying twenty-five missions as an F-5 Lightning pilot with the 7th Photo Reconnaissance Group based at Mount Farm in Oxfordshire. The latter's statement on the actual flying conditions on that fateful day greatly assisted the crash investigators and the board's recommendations. The crew members on Bloemendal's B-24 who also died in the accident were Sergeant Gordon Wilbur Kinney and Corporal James Manuel Parr.

A memorial fund was set up after the Freckleton crash and it was decided that the most fitting tribute would be in the form of a children's playground, and arrangements were made to purchase the necessary land. Donations from the Warton servicemen were entirely voluntary, but every person on the base, without exception, made a contribution. The playground, now used for over fifty years, is always visited by the B.A.D. 2 veterans who come 'over to home'.

BAD 2 hangars and ramps in September 1944.

The World's Greatest Air Depot

This is what it's all about! **Shack Time**, the B-24J 44–40275 of the 458th Bomb Group, at Horsham St Faith near Norwich is on a mission to Germany. Originally equipped to operate the Azon controllable bomb, the aircraft had its special aerials removed after visiting Warton in September 1944. Unfortunately, it was lost in a crash two months later on 14 November.

Chapter 7

Returning to Normality

Every man and woman at Warton was affected by the Freckleton crash, but it was evident that the disaster could not be allowed to affect the station's performance. Luckily, a visit by Bing Crosby and a number of entertainers had been scheduled for 1 September 1944, and as this went ahead it did serve to alleviate some of the sadness which prevailed. Bing and his troupe served up some excellent entertainment from a makeshift stage in Hangar 7, which was packed with an overspill into Hangar 6. His songs included 'White Christmas', 'San Fernando Valley', 'Easter Parade' and more of his favourites. All the entertainers took time to sign autographs and spent time talking to the Wartoneers. Bing visited the base after only a few days in England, and soon after leaving Warton he went to France to cheer the troops at the front. The station personnel considered themselves quite fortunate to be present at one of the few shows he gave in England. Because the visit was so soon after the tragedy, Bing insisted on going to the base hospital to entertain the survivors. Although it delayed his departure he stayed at the hospital long enough to visit every patient, and sang two songs for the children, some of whom were terribly burned.

Another morale-boosting visit occurred on 13 September when the boxer Billy Conn kept his promise to return to Warton for some exhibition bouts. In July he had visited with world champion Joe Louis, but this time he

On 1 September 1944 Bing Crosby visited Warton to perform for the troops, but he also found time to call in at the base hospital to sing for the survivors of the air crash. This snapshot of the 'Old Groaner' was taken by Les Smith, a young R.A.F. airman who was at Warton training to service American equipment.

Some of the lovely ladies of the American Red Cross pose for the base photographer making a very pleasant change for him rather than his everyday 'nuts and bolts' photography.

The B-17 transport 42–97108 was originally HB771, an R.A.F. Fortress III which was transferred to the U.S.A.A.F. for use by the VIIIth A.F.S.C. The aircraft is unusual in having invasion stripes applied as they were not normally seen on the heavy bomber types.

This 'Carpetbagger' decided to take to the grass after a brake failure on landing.

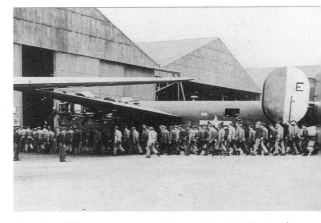

Another shift commences at Hangar 4 on 20 September 1944. The B-24 wears the markings of the 34th Bomb Group which was at that time in the process of converting to B-17s.

Returning to Normality

Lieutenant-Colonel A.W. Nelson, Warton's Station Commander, poses with test pilot Pete Swank after a flight to southern England. Colonel Nelson, who had flown Sopwith Camels during World War One, wore the wings of a command pilot.

The B.A.D. 2 ramp consisting mainly of new Mustangs, although two aircraft in the rear row are from the 363rd Fighter Group of the 9th Air Force. These machines had arrived for modifications after that group had changed to the tactical reconnaissance role.

Above:
The American-operated Airspeed Oxford **Sad Sack** was a regular visitor to Warton. The aircraft had been painted silver, although most of this type retained their British camouflage or had received the U.S.A.A.F.'s olive-drab colour scheme.

Left:
Fay Brandis, centre, the regular pilot of the B-26 **Demon Deacon**, with Major Norman Isenberg, Commander of Site 13 (left), and Captain Bob Forehand, an engineer. This photograph provides an excellent illustration of the Marauder's nose art.

The World's Greatest Air Depot

One of the first A-20 Invaders to pass through Warton was 41–39222, seen here as Peter Manaserro flies the aircraft over the outskirts of Preston while en route to base.

came alone and, after Special Services had erected a ring in Hangar 7, Conn gave the boxers among the base personnel a couple of rounds of sparring.

The Mayor of Blackpool sent an invitation to Colonel Moore for a contingent of troops from Warton to take part in a parade to mark the fourth anniversary of the Battle of Britain. On Sunday, 17 September, men of B.A.D. 2 joined the large parade which marched through the town and were later complimented by the mayor on their smart appearance and deportment. Colonel Moore had been delighted to receive the invitation, but was even more so to receive a letter from the mayor congratulating him on the fine showing of his men.

There was some sadness on both sides as the 87th Air Transport Squadron departed Warton after being ordered to move to A.A.F. Station 519 at Grove in Berkshire. Good friends had been made and the move, while not unexpected due to the action on the Continent, did cause a certain amount of

The ever popular Colonel Paul Jackson was on hand to open the Warton Hobby Shop and, needless to say, the favourite hobby was aeromodelling!

*The blue-nosed P-51D **Nancy M 2nd** of the 352nd Fighter Group arrived for overhaul from its base at Bodney in Norfolk.*

You're holding up the war! This contractor's steamroller trundled along in front of the B-24 for more than ten minutes before it turned off the perimeter track to let the bomber pass.

Above:
Dopey Gal, *coded C3-B, of the 363rd Fighter Group waits with other Mustangs to be rolled into Hangar 5.*

Below:
It was a bad day on 5 October 1944 for Warton's accident statistics as three aircraft came to grief. The first was when the B-24J 44–40466, 6L-L of the 466th Bomb Group, ran out of runway.

Returning to Normality

The B-24 crash was followed shortly afterwards by the A-20K 44–561 which went even further off the runway while trying to avoid the Liberator.

The dreaded Runway 15/33 claimed its third victim when a new P-51D arriving from Speke bent itself rather badly when trying to miss the other two aircraft.

Undercarriage retraction tests in Hangar 4 were supervised by Staff Sergeant Tony Piscatelli who, in later years, became Chief of Police for Elizabethtown, Pennsylvania.

depression. The transfer was completed by 18 September.

More sadness followed on the 25th when it was announced that one of the popular newcomers, Private Edward Farow of the Maintenance Division, had died of his injuries after walking into the spinning propeller of a P-38 during the night shift.

On the same day, a new base commander, as opposed to the depot commander, arrived in the shape of Lieutenant-Colonel A.W. Nelson. The former commander Colonel Robert G. Patterson, who had held the position since the U.S.A.A.F. took over, was transferred on 23 August 1944 to the 1st Allied Airborne Division, and since that date Major Ben Goodman had been acting as base commander. Lieutenant-Colonel Nelson arrived from A.A.F. Station 597, B.A.D. 3 Langford Lodge. He was a veteran of the First World War and was a command pilot.

Production figures had slowed down slightly although this was not through a lack of effort: straightforward maintenance and overhaul work was continuing although time was now being taken up by the variety of modifications needed on a number of aircraft types. A number of P-51 Mustangs were being converted to the F-6 version for the photographic reconnaissance role while B-24s arrived to be modified for duties as weather, night leaflet and fuel-carrying aircraft. Two B-17s arrived for conversion to weather aircraft while the usual kits for the assembly of Piper L-4 Grasshoppers continued to roll in.

Another new aircraft type arrived at Warton on 4 September 1944 when Captain Fay Brandis ferried in the Douglas A-26 Invader 41–39203. The first aircraft stood alone on the ramp for four days before test pilot Jack Knight called on flight engineer Spence Thwaites to join him in a test flight. The conversation between the two went something like this: Knight: 'Let's go fly.' Thwaites: 'Have you got the pilot's notes?' Knight: 'Nope.' Thwaites: 'Have you got any paperwork?' Knight: 'Nope.' Thwaites: 'Do you know how to fly it?' Knight: 'I don't even know how to get in it!' Such was the pressure

The B-24L 44-49359, converted to a transport aircraft, wears the code letters of the 32nd Squadron of the 314th Troop Carrier Group of the 9th Air Force, although no records show that the unit operated a Liberator. The Lancaster behind belonged to the R.A.F.'s No. 5 Lancaster Finishing School.

The night shift working in Hangar 4. Some men liked to work nights, but whatever the shift every man knew the operation done by the previous team and the work continued uninterrupted.

Captain Fay Brandis, right, had a pleasant surprise when he delivered a P-51D from Warton to the 361st Fighter Group at Little Walden in Essex, for while he was there he met fighter pilot Bill Quinn who was a school classmate!

of war at B.A.D. 2!

Besides the scheduled duties, many unofficial modifications were carried out, and one such came when C-47 pilots of the Troop Carrier Command complained about the lack of protection against small-arms fire while on low-level parachute drops. Captain Royal G. Thern, along with Captain Smith of the Aircraft Salvage Department, arranged for armour-plated sections recovered from scrapped B-24s to be cut down and shaped as back and foot plates for the C-47 pilots. It came as a great pleasure to everyone involved when Colonel Moore received a memo from T.C.C. Headquarters telling of how the seat plates had saved the lives of the pilots of a C-47. It was reported that the aircraft was taking off and towing two CG-4 gliders. The gliders were off the ground, but the transport was still on the runway when its undercarriage collapsed with the left gear folding up first, and as the port propeller hit the concrete a large piece of the blade broke off and sliced through the nose of the C-47. It struck the pilot's seat a terrific blow and then hit the co-pilot's chair smashing the arm. Surveying the

The last Allison V-1710 engine used in the P-38 Lightning was tested by the Engine Repair Department on 9 October 1944. It was the 2,586th of its type to be processed by the department which would, after that time, concentrate on the Packard Merlin V-1650 which powered the Mustang.

One of the many Anglo-American weddings took place at St John's Church in the centre of Blackpool on 7 October 1944 when B.A.D. 2's Harry Goldsmith married Miss Betty Holmes. Even now the couple, who live in California, look much the same although Harry hasn't quite as much hair these days.

damage afterwards, the armour plating had undoubtedly saved the aircraft captain's life and neither pilot was injured. In similar instances before the fitting of the armour plating, one pilot had been killed while another lost a leg. Colonel Moore had copies of the memo posted all around Warton and expressed his pleasure and pride to receive such a communication as it was a tribute to B.A.D. 2 and especially to those who had devised and fitted the modification.

Following the urgent move of the 87th A.T.S. to Grove, the 312th Ferrying Squadron, which had arrived at Warton from Langford Lodge, was also ordered to Grove and departed on 27 September 1944.

The B.A.D. 2 football team, known as the 'Warriors', played its first game on 1 October when it beat the Burtonwood 'Bearcats' 6–0 at Bloomfield Road, home of Blackpool Football Club. A crowd of 30,000 watched the game for free, but a collection raised £500 which was presented to the R.A.F. Benefit Fund by Major George Ashman of Warton's Special Services. The 'Warriors' went on to become the undisputed champions of American Football in the E.T.O., having never lost a game!

A memorable occasion took place on 5 October 1944 with the first all-American wedding at Warton. The bride, First Lieutenant Ellen Holcomb of the U.S. Army Nursing Service, was given away by Colonel Moore when she married First Lieutenant Willard Wildbur, an engineering officer at B.A.D. 2. The couple did not want to be married on the base so the wedding was conducted by the Reverend Ronald Halstead at Warton Parish Church.

While the ceremony was taking place

*The B-17G **Flak Hopper** had its gun turrets removed for the conversion to a transport. These conversions were popular with pilots as the weight saving made them into 'hot ships'.*

another unforgettable incident was occurring on the airfield when there were three aircraft accidents within fifteen minutes of each other. The runway 15/33, which was 3.960 feet (1,320 yards) long, was seldom used for operations and normally acted as a parking area for the overflow of aircraft. However, on 5 October strong winds brought the runway into use with the aircraft approaching over the River Ribble to land in a north-north-west direction. Where it intersected with the main runway a depression had appeared, presumably with the latter settling. Because of the dip, many aircraft became airborne again once they had touched down and an unsuspecting pilot would suddenly have his hands full trying to get the aeroplane down in time. The first aircraft to come to grief was a B-24 from the 466th Bomb Group at Attlebridge in Norfolk which, after passing over the main runway, became airborne again, and after touching down for a second time, ran out of runway. After a squeal of brakes the bomber ran off the end of the concrete and into the soft grass, ripping off the nose wheel. However, Warton was a busy airfield and a little incident like this would not be allowed to stop the flying. Just twelve minutes later an A-20 returned from a test flight and exactly the same thing happened, although this time the Havoc pilot had not only to stop, but to steer his aircraft around the Liberator sitting tail high in the grass. Through skill and not a small amount of luck he was able to veer off on to the grass, missing the B-24. Despite the two aircraft in the grass at the end of the runway and not too far from Hangar 7, it was thought safe to let a P-51 land. Once again an aeroplane became airborne after passing over the depression with the Mustang leaping into the air, and before the pilot could react to the problem the aircraft was on the

*Technical Sergeant Ralph Scott of the Radio Section is now the editor of the B.A.D. 2 Association's excellent magazine, aptly named **BAD News**.*

runway racing towards the crashed bombers. With much power being applied the Merlin's torque pulled the P-51 off the runway and into the grass where it ground-looped. The pilot was able to climb out through a split in the fuselage and soon joined the other aircrews in the base hospital. Luckily, the only thing hurt was their pride. It was later concluded that the depression and the strong wind had been the main factors in the accidents.

It was a busy day on the 5th because the whole of the 40th Air Depot Group moved from Warton to A.A.F. Station 169 at Stanstead in Essex. On the same day the 1st Air Service Group was formed incorporating the 519th and 520th Service Squadrons and the 1988th QM Truck Company (Avn). Six days later the 1st Air Service Group was transferred to A.A.F. Station 385, Halesworth, Suffolk.

On 7 October 1944 more transfers took place when one officer and twenty-four enlisted men of the Maintenance Division were temporarily assigned to A.A.F. Station 519 at Grove, Berkshire, to assemble a number of Piper L-4 Grasshopper liaison aircraft. One week later the 310th Ferrying Squadron followed them to Grove, although seventy-five pilots were left behind to join the Flight Test Department of B.A.D. 2. It was thought that this transfer would establish a better system for the delivery of aircraft as the increasing production rate required a more urgent approach and, in some cases, the machines had to be flown to forward airfields.

The ingenuity of Warton's technicians in devising methods to improve efficiency was endless, with working times slashed on every type of aircraft. An excellent example was when Staff Sergeant Paul E. Smith invented a magneto wrench for the inspection of the Northeast magnetos on the P-51s. His design enabled mechanics to remove the rear casting screws without having to remove the complete magneto from the engine, thereby saving over two hours per magneto or four hours per aircraft. The original method was difficult, time-consuming and also produced many skinned knuckles.

In the Engine Repair Department the last Allison V-1710 engine was tested on the 9th, the 2,586th of its type. When the facility was established, British 'W' type engine test stands were converted by the technicians to run both the V-1710 and the V-1650, but initially trouble was encountered as the engines ran too cool. Two bypass valves were devised and this corrected the problem with the added bonus of decreasing the time on the whole test procedure.

The great requirement for engines during 1944 saw the section expand into Hangar 14, which had been used by the Salvage Department. This move allowed better facilities for engine breakdown and cleaning with the production lines being rearranged until they were operating efficiently. In May further expansion saw the engine people take over Hangar 15, formerly home to the Dispatch Department. The steady increase in production of engines cleared for service had brought a commendation for all in the section from B.A.D.A. Headquarters, but on 9 October 1944 another communication from headquarters directed that production of the V-1710 should cease with immediate effect. Production was halted on the Allison with the exception of a few engines which were having small modifications, but the department

Looking extremely smart in their white silk scarves are the officers of Flight Test: (L to R) Dewey Campbell; Mike Murtha; Ed Sayers; Pete Swank; Charlie Himes; Bill Watters; Adam Talaki; Jack Knight; Wally Woltemath; Joe Bosworth; Peter Manaserro; and Thompson Boland.

would continue to turn out the V-1650 as this engine would be unaffected by the order.

In Warton's Bombsight Department a directive had been received showing concern for the stabilising mount of the Norden B-7 bombsight, but the solution came from B.A.D. 2's specialists who modified a template designed by the manufacturers to cure the problem. The Norden section also constructed a C-1 autopilot mock-up which enabled the line crews to complete on-the-spot troubleshooting, and published instructions for engaging and adjusting the autopilot during air tests.

The Base Hospital, under the command of Lieutenant-Colonel Julius L. Sandhaus, had completed numerous additions and improvements with nine wards and an annexe being added during 1944. A total of two hundred and ten beds were available, and with the nature of Warton's activities plus the ever-increasing number of personnel the hospital was kept reasonably busy. However, on 27 October Second Lieutenant Edward F. Honig took command in an arrangement which could allow Dr Sandhaus to devote all his time to his duties as a surgeon.

The 56th Field Hospital had come under the jurisdiction of the station's hospital, but on 28 October the airfield accommodated forty-six visiting C-47s which had arrived to transfer the whole of the unit to Le Bouscat, France. The operation ran smoothly with the additional movements causing no problems to the highly experienced Warton air traffic controllers.

On 6 November General Knerr C.G. A.F.S.C. paid a surprise flying visit to Warton and was most impressed by the professionalism of the personnel and cleanliness of the base, particularly as the tour of the facility was entirely unexpected. The following day there was a scheduled V.I.P. visit by General Ott C.G. B.A.D.A. who arrived for a medal award ceremony. Two officers and three enlisted men were receiving the Bronze Star, while Distinguished Service Awards were presented to fifty-two of the B.A.D. 2 personnel.

CHAPTER 8
Change of Command

It was a sad day for Warton on 9 November 1944 when word was received that the Commanding Officer, Colonel John G. Moore, was to be relieved for an assignment at U.S.A.A.F. Headquarters in Washington D.C. Colonel Moore had taken command of B.A.D. 2 on 29 October 1943 and it was under his able direction that the depot had achieved the outstanding record of production. On the 20th Colonel Moore carried out his last official duty when he awarded a citation to the Instrument Department when the 200,000th instrument repaired was presented to the Supply Division. He made a brief address to the men before he himself was presented with a gyro-compass which had been suitably mounted with an inscription by the Instrument Department. The presentation was made by the departmental head, Captain Hacker, but the tables were turned when the popular

Colonel Don L. Hutchins at his desk after assuming command of B.A.D. 2 on 25 November 1944.

Captain was presented with his Major's oak leaves by Colonel Moore. The much surprised ex-Captain left the ceremony close to tears.

As a gesture of appreciation in bidding farewell to Colonel Moore, the Warton officers gave a party for him at the Winter Gardens in Blackpool that evening. Officials both military and civil attended; among them was Major-General Earle E. Partridge, C.G. 3rd Bombardment Division, Brigadier-General Clarence P. Kane, Deputy C.G. A.F.S.C., General Ott, and the mayors of all the local towns. Presented to the guest was Colonel Don L. Hutchins who would be taking command as of 25 November 1944. Colonel Moore remained at Warton until the 30th meeting as many of the men as possible. He had earned the respect of everyone during his period at the base, not only with his leadership but by his personality and his

Colonel John Moore, left, in conversation with his successor Colonel Don Hutchins during the former's farewell party on 24 November 1944.

The World's Greatest Air Depot

Above:
In November 1944 a war-weary C-47 named **Chukky** was declared surplus to requirements, but arrived in Hangar 4 to be rebuilt in a similar fashion to the P-51 **Spare Parts**.

Below:
Chukky, alias 41–38607, received some tender loving care from B.A.D. 2's engineers and emerged as the beautiful **Jackpot**, named for Colonel Paul Jackson.

Change of Command

Above:
An excellent side view of the now gleaming **Jackpot** *after it was towed out on to the flight-line for its first flight after the rebuild.*

Below:
Jackpot *became a familiar sight at Warton and survived the war to give many years of fine service flying around Africa. The AT-6D in the picture is B.A.D. 2's own* **Jet Threat**, *42–85163.*

knowledge and interest in those under his command. He departed for the United States on the morning of the 30th, but was still busy visiting sections and saying his farewells until the take-off time.

In mid-November 1944 a war-weary C-47, 41-38607 *Chukky*, was wheeled into Hangar 4 for a complete rebuild. The aircraft had been declared surplus by the IXth Troop Carrier Command and instead of going for scrap it was decided that the Wartoneers would rebuild the transport for use as a station hack. The olive-drab machine was completely stripped down, worn-out parts were replaced, and a number of additions were incorporated to make the aircraft 'that little bit special'. As the now gleaming natural-metal C-47 was ready for roll-out the name *Jackpot* was given to the machine as a tribute to Colonel Paul Jackson, Chief of Maintenance. 'Jackpot' had been the Colonel's call-sign while flying from Warton, and as this experienced officer was one of the pilots responsible for introducing the Douglas C-47 into military service in 1941, it seemed an appropriate name. This aircraft served the 'Warton Air Force' well, and after many morale-boosting 'supply' flights the old C-47 was flown by Colonel Jackson on many occasions to fly B.A.D. 2 personnel on trips to Europe after hostilities had ended. The Hangar 4 people must have done a good job on the old bird because after the war it was bought by Trans-Atlantique, a French cargo operator, as F-BEFH before being sold to Air Cameroun. After many years of operating in Africa old *Jackpot* was last heard of in storage near Paris in 1971. With the number of DC-3/C-47 types still flying in many parts of the world it would not be a surprise to hear that she is one of them.

Just four days after Colonel Hutchins had taken command of B.A.D. 2 the station had its first serious aircraft accident since the crash of the B-24 at Freckleton. Two of the new Douglas A-26 Invaders collided after take-off for delivery to the 409th Bomb Group of the IXth Bomber Command based at Bretigny in France. In September the first four of the A-26s were cleared for delivery from Warton and these were followed in October by a further eight. By November the B.A.D. 2 engineers had really gone into overdrive because one hundred and fifteen Invaders were completed and readied for delivery, but two aircraft did not complete the delivery. The first of the type to arrive at Warton were A-26Bs with the armament alone causing a few gasps among the station's aviators and ground crews alike. The aircraft had an 'attack' nose containing six machine-guns with remotely controlled dorsal and ventral turrets each containing two machine-guns. This armament was supplemented by eight more in packs under the wings, and with all these guns being fifty-calibre the A-26 had more than twice the firepower of a P-47 Thunderbolt. With an additional bomb load of up to 4,000lb it made the Invader a potent aeroplane. There was also provision for replacing the nose guns with one 75mm cannon, but no such modification was carried out at B.A.D. 2. The A-26 was a new aircraft with which Warton's technicians had to quickly acquaint themselves, and it was generally well accepted, although it was prone to nose-wheel collapse. The two aircraft involved in the accident were flown by Lieutenants Hubbard and Zuber, serialled 43-22298 and 43-22336. At approximately 1215 hours on 29 November 1944 the Invaders collided as they circled the airfield after take-off, with both machines falling into the River Ribble estuary. A first-hand account was given by Sergeant Stanley C. Begonsky of the Maintenance Division:

> I was near the Salvage Hangar about 1210 hours when I looked up and saw two airplanes collide and plummet into the Lytham channel. I got out on to the road and an M.P. jeep picked me up and took me to Site 8. We stopped there and I ran across the field to the water's edge. The tide was in and the water was deep. One of the planes was smoking kind of bad and the other was sticking straight down. Someone on the bank yelled that they thought that there was someone alive on the tail of the burning

The rows of A-26 Invaders awaiting collection stretched around Warton's perimeter track. Most deliveries went smoothly until 29 November 1944 when two of the aircraft collided soon after take-off, with fatal results.

The never ending lines of A-26s duped the author into thinking that the aircraft were based at Warton as he did not realise they were constantly changing. New aircraft were arriving daily for processing to take the place of those delivered to 9th Air Force groups.

plane. I waded through some marshes as far as I could and from there I swam over. I swam about a hundred feet through the swift current to the plane with my heavy flying clothes on. I was the first one there. There was a dinghy coming towards the wreck from way off. The dinghy never made it, but some English eventually arrived in a rowboat. When they came they looked at me and wanted to know how I got there. They thought I was one of the survivors. I looked in the plane which was not on fire and found the pilot and co-pilot dead.

The latter was not a pilot, but Private John F. Guy who was a passenger on the flight. I left them and waded over to the burning plane which was approximately seventy-five feet away. Smoke was belching so bad I couldn't see anything. I crouched down on my haunches and got up close to the plane and chopped out a portion of it. I could see the pilot in his bucket seat and the flames were licking up the back of his jacket. I could then see that his pants and parachute were burning. The only way I could get him out was by chopping off his parachute and dragging him out. His legs were broken and the sharp edges of the plane were sticking in him. I dragged him out some of the way. The smoke was getting so bad I had to give way for a while. I got back and finally got him out. I dragged him about twenty-five or thirty feet from the flames. He didn't have much clothing on at that time. I had to stop once as I was dragging him out because I got on fire myself. I went back in to see if there was anybody else there as I thought there were two. I looked back in and couldn't see anything and the flames were beginning to die down. The British Air/Sea Rescue [the Lytham St Annes lifeboat] had arrived and we searched the plane thoroughly and could not see anything. I told the British we should try to get the bodies ashore and I cut strands off the parachute to cover them up. I got a dinghy, inflated it, and placed the bodies in it then dragged it to the boat.

An Irish vessel making its way up the Ribble to Preston Docks makes an interesting picture, but mainly because of Warton in the background with A-26s and B-24s parked as far as the eye can see.

A Police War Reservist, Harry Crompton, also swam to assist Sergeant Begonsky before the lifeboat arrived, and later the bodies were landed at the boarding stage at Lytham. The Royal National Lifeboat Institute recorded that, 'Sergeant Begonsky had gallantly swum through the dangerous channel to try to assist the airmen and deserves the highest praise'. Although not in the Sergeant's report as he and the lifeboat men were at the burning aircraft releasing the two bodies, ammunition and flares were exploding making the task even more hazardous. The rear fuselage and tail of 43–22298 remain to this day buried in the mud where the aircraft fell, but brave attempts by aviation archaeologists to remove it are always frustrated by the tide which covers the area after great progress has been made. The aircraft's fin still appears above the

David Stansfield led members of the Pennine Aviation Museum in an attempt to recover the rear fuselage of the A-26 43–22298 which was involved in the mid-air collision on 29 November 1944. Unfortunately, at the time of the dig, nearly forty years after the crash, the tide frustrated all efforts and the aircraft still lies where it fell.

A Liberator transport lands on Runway 08 after a test flight piloted by Peter Manaserro.

Working out of doors was great in the warmer weather, but not too pleasant when the cold damp winters set in.

mud and this has frequently been reported as a 'Flying Fortress' or a 'Mustang', and practically every other type that flew from Warton.

In the production records of B.A.D. 2 there is no record of C-109 Liberator tankers and U.S.A.F. records show that this type was only used for carrying fuel to advanced bases in China to serve the B-29 units operating against Japan. Although reported as such, the fuel tanker conversions carried out at B.A.D. 2 were not classed as the C-109, but remained B-24s. Four bomb groups of the 2nd Air Division had the task of having some of their aircraft converted and these included the 458th, 466th, 467th and the 492nd, who were given the role of hauling fuel and supplies to General Patton's 3rd Army and the 12th Army Group fighting on the Continent. Although the fitting of the fuel tanks to the B-24s was also carried out at other air depots,

A P-38 provided a fine platform for a group photograph of the Flight Test team. Under the nose are Charlies Himes and Tom Boland with Jack Knight sitting at the front. Standing on the extreme right is Spence Thwaites, past president of the B.A.D. 2 Association, with Wally Woltemath next to him. Other B.A.D. 2 Association members in the shot include Orville Wrosch, Ray Dlouhy and the late Frank Segalle. Apologies to any forgotten members!

*One of the last known photographs of **Spare Parts** as Joe Bosworth flies it over the uninviting mountains of North Wales.*

Some of the 'Warton Air Force' photographed on Runway 15 with the UC-78, P-47, C-47, P-51 and AT-6 in attendance.

Above:
As no mission markers or victory symbols appear under the cockpit of the P-38J 44-23152, it would seem that the Lightning was living up to its name.

Below:
The 'Warton Air Force' people. (Rear row, L to R) Andy Barnes; Edward Sayers; Dewey Campbell; Bill Watters; Wally Woltemath; Joe Bosworth; and Pete Swank. (Front row, L to R) Ken Stroehle; Dave Mayor, founder of the B.A.D. 2 Association; Paul Jackson; Charlie Himes; Mike Murtha; and Peter Manaserro.

Some of Warton's unsung heroes, the fuel tanker drivers and Alert Crew of B.A.D. 2.

the bulk of the work was allocated to Warton.

Although stationed in an attractive area and near to the country's most famous holiday resort, life was not all play for the men of B.A.D. 2. As production built up so did the required working hours for the technicians. In July 1943 hangar personnel worked eight-hour shifts, and in the following November these were increased by one hour. Just one month later ten-hour shifts were introduced and by February 1944 the shifts were equally rotated as day and night with all work continuing around the clock. The stress of such working with only occasional time off was beginning to tell on the health of some of the men, and in May, on medical advice, the working schedule was revised to allow all personnel to have one day off per week. In September 1944 the men received their first leave, including passes for three, four or seven days provided no more than five per cent of any section were absent at the same time. Besides the attractions of Blackpool, Manchester or Liverpool, many of the men headed towards Scotland. The following month the shift system returned to eight hours much to the delight of the men, although there was no reduction in the output of aircraft. The teamwork and efficiency of everyone involved brought about the change.

Although titled the Bombsight Department, this section was completing work on autopilots in equal proportions to the bombsights. In early November the department accomplished the installation of an electronic C-1 autopilot into a C-47. All the wiring diagrams and installation plans were

The 1,000th aircraft to be processed in Hangar 4 was rolled out on 15 December 1944. This B-24L, 44–49591, cleared its flight tests and was delivered to the 467th Bomb Group at Rackheath in Norfolk just a few days later.

B.A.D. 2 operated this Norseman early in 1945, and although it was a noisy aeroplane it had a large carrying capacity and was popular with the pilots.

devised by the men of the department as the work had never been carried out before. The old A-3 autopilot was removed and the newer, more efficient C-1 installed, and the aircraft successfully flight tested all in less than one week. Reports on the success were sent to the Douglas Aircraft Corporation and the C-1 was recommended as an alternative autopilot as a result.

The departure of the Ferry Squadron left the handsome terminal building vacant and this was taken over as headquarters of the Flight Test Department. The building was completely changed around with offices built for all the section heads, and with the operations section included in the facility the building was redesignated the Air Terminal Building. Under Major Gale E. Schooling, the department comprised five new sections with the first including Flight Surgeon, Accident Investigation, Passenger and Freight Services, Terminal Manager and the Link Trainer; the other sections were Flight Test, Flight Test Engineering, Aircraft Delivery and Flight Test Operations. Major Himes established a Pilot's Transition Training School for the purpose of training and checking out all pilots in every type of American aircraft flown in the E.T.O. The Aircraft Delivery section was one of the new organisations which had to start from scratch as the pilots and crews could now be assigned to delivering aircraft. This role was soon experienced when test pilots Knight and Manassero were ordered to deliver a B-24 to Italy.

The modification of aircraft types became a standard part of B.A.D. 2's working programme, and besides the increasing demands of the 'Carpetbagger' project, others such as the night leaflet aircraft were now joined by the requirement of B-24s specifically for flare dropping, a project known as 'Firefly Big'. A similar project was then added to the modification programme when A-20 Havocs were converted as 'Firefly Little'. Another

The World's Greatest Air Depot

A cold day in January 1945 with two snow-covered B-17s parked near the River Ribble. The far aircraft, with its gun turrets removed, was ex-VK-B of the 303rd Bomb Group.

requirement was for A-26s to be equipped with three signal lights on each wingtip and, once again, the B.A.D. 2 technicians had to design and fit a system to meet the requirement. A strange request came from A.F.S.C. Headquarters with the requirement for flak repair kits to be manufactured as linemen in the combat groups spent too much time producing them. Warton then devised a system to make the patches in six different sizes and proceeded to turn out 10,000 a day.

On 15 December 1944 it was a memorable day for Hangar 4 as its 1,000th aircraft, the B-24L 44–49591, was completed and rolled out. The record showed that altogether ninety-two men had worked on the aircraft during its modification and inspection. Colonel Jackson was on hand to receive the Liberator in a short ceremony held at the hangar door.

As pressure to produce the combat types increased once more, the Engine Repair Department was required to allocate some men for the assembly of L-4 and L-5 liaison aircraft with effect from 16 December. The facilities at Hangar 15 were made available for this work, with the Dispatch Department having to relocate to Hangar 13 in the space formerly occupied by the Allison final assembly line. This extra production of liaison types commenced on 18 December with the first aircraft, the Stinson L-5B 44–16902, coming off the line just two days later.

A communication on 5 December had caused great concern to Colonel Hutchins as the recommendations of the Aircraft Accident Board, regarding the B-24 crash at Freckleton, stated that the Station Flying Control Officer and the Station Operations Officer should be one and the same. After careful study of the document, Colonel Hutchins telephoned Colonel Milo McCune, Deputy Commander B.A.D.A., on the 19th to tell him that B.A.D. 2 would not undertake to put the recommendations in place. The Base would appoint a flying officer in the full capacity of the Operations Officer while the Flying Control Officer would be a ground officer having graduated from Flying Control School and would perform only those duties for which he was trained. Colonel McCune, somewhat taken aback, replied that he would

Change of Command

It was a lonely vigil for the air traffic controller in his van parked near the end of the active runway. In the winter months the lack of heating was the main problem until Warton's engineers devised a heater.

investigate the matter with other units if Warton felt so strongly about the recommendations.

Another cause of concern for Colonel Hutchins came as he was advised that some of the German prisoners recently placed in a mill at Kirkham were flying personnel, and in the event of a break-out immediate action should be taken to prevent aircraft based at Warton being used to make good their escape. British troops were available to assist Warton if required. This concern was increased on Christmas Eve when the Colonel received a telephone call at 1630 hours stating that a plot had been uncovered whereby the 28,000 German prisoners within a thirty-miles radius of Preston were planning a mass escape, the object being to sabotage as many military installations as possible. The Base was put on full alert and personnel were armed, but three days later the emergency was scaled down and held in abeyance for reintroduction at a moment's notice. It was also noted that at the time of the alert there were eight hundred and thirty-eight aircraft on the airfield at Warton.

With the incredible facts and figures required to operate B.A.D. 2 it is interesting to note that by the end of the year the Supply Division received 50,000 tons of supplies and shipped approximately 45,000 tons of completed items. An average of two hundred and ninety-three trucks were dispatched weekly, hauling supplies all over the United Kingdom. One hundred and forty railway wagons were incoming while one hundred and twenty-three were shipped out. During 1944 302,482 items of varying quantities were requested and 61,000 teletypes received.

El Diablo, one of the small tail B-26 Maurauders seen over a snow-covered Lancashire in January 1945. The aircraft had served with the 322nd Bomb Group of the 9th Air Force before becoming a training aircraft for the 3rd Combat Crew Replacement Centre. Its code letters, partly obscured by the tailplane's shadow, were W9-I.

Chapter 9

Victory in Sight

The new year of 1945 started on a high note for B.A.D. 2 as on the previous day Warton's football team, the 'Warriors', had been proclaimed champions of the European Theatre of Operations after more than 25,000 people saw them beat the 8th Air Force 'Shuttle-Raiders' 13–0 at the White City Stadium in London.

However, the euphoria disappeared on 2 January 1945 when Warton received a telephone call from Lieutenant Courtland Crandell to report an aircraft accident near Slaidburn in Lancashire. Ambulances, crash trucks and a section of military police were immediately sent to the scene. The aircraft was the B-24 Liberator 42–100322 of the 448th Bomb Group at Seething in Norfolk which was on its way to Warton for an overhaul. There were seventeen men on board, which included a double crew to take two completed B-24s back to Seething and nine servicemen who were taking their leave in Blackpool and were travelling in the Liberator. The old aircraft was having trouble with its navigation aids and during a descent in cloud the machine had struck the top of Burn Fell, a hillside two miles from Slaidburn; the bomber slid across the moor and only came to a halt when the rear fuselage swung around into a stone wall. Four fatalities were recorded, all were located in the part of the fuselage which impacted with the wall. Crandell later turned out to be the pilot.

One day later came more bad news when it was learned that Flight Officer Edward Johnston had been killed in an aircraft accident at Little Walden in Essex while delivering a P-51 to the 361st Fighter Group. Johnston was serving with the 310th Ferrying Squadron, but was on detachment to Flight Test at Warton.

An excellent landing shot of **Donna-mite**, *a P-51K of the 353rd Fighter Group, returning to its base at Raydon in Essex after a mission. This aircraft, which passed through Warton in December 1944, was one of the first of its type in Britain. The difference from the P-51D was down to a change in the propeller from Hamilton Standard to Aeroproducts.*

The Aeroclub was always popular with the Wartoneers. Run by the American Red Cross it provided a little touch of home.

Satan's Daughter, a Douglas A-20, had completed seventy missions when it was returned to Warton. It was eventually flown to Langford Lodge in Northern Ireland for disposal.

A very heavy landing and subsequent undercarriage collapse pushed the A-20's main wheels up through the engine nacelles. The aircraft had been ferried to Warton for disposal, but the pilot seemed to have done half the job of the Salvage Department.

The Liberator 44–49591, the 1,000th aircraft to pass through Hangar 4, just about to touch down at Rackheath in Norfolk, home of the 467th Bomb Group. This group, along with the 392nd, painted the last three digits of the aircraft's serial number on both sides of the nose.

As the land forces on the Continent were beginning to be stretched on many fronts, an order came from Supreme Headquarters that a number of enlisted men should be put forward for infantry combat training. Staff Sergeant Joseph L. Gaudette of the Administrative Division voluntarily took a reduction in rank to Sergeant in order to get on the combat shipment. It was an admirable gesture which won plaudits from his superior officers, although most of his colleagues thought him to be a little misguided.

On 11 January 1945 yet another accident brought a sombre attitude to Warton when the R.A.F. at Calverley in Cheshire informed B.A.D. 2 that a P-51 Mustang had crashed near Spurstow killing the pilot. The aircraft, flown by First Lieutenant Leonard D. Johnson of the Delivery Section of the Maintenance Division, was en route to Debden in Essex on delivery to the famous 4th Fighter Group.

Colonel Hutchins played host to a number of senior R.A.F. officers on 17 January and the B.A.D. 2 aviators were enthralled by talking to Air Vice-Marshal Raymond Collishaw who was a fighter ace during the First World War.

During January the last of the replacement pilot seats were fitted to B-24s. The Liberators were originally fitted with armoured seats for the two pilots, but while these afforded excellent protection from flak fragments, the seats were cumbersome, restricting movement and, in the case of emergency, a quick exit from the aircraft. Known as 'coffin seats' to the B-24 crews, B.A.D. 2 started changing these in May 1944, and as new aircraft were arriving from the manufacturers with improved seats, the replacement programme came to an end. External armour too was fitted to fuselage sides under the cockpit windows of B-24s, much to the relief of the Liberator pilots, and this too was discontinued as new aircraft arrived.

An interesting task in January was the work carried out on eight P-38 Lightnings which arrived at Warton from A.A.F. Station 237 at Greencastle in Northern Ireland. Greencastle, a satellite of B.A.D. 3 at Langford Lodge, had classed the aircraft as war-weary and sent them to B.A.D. 2 for disposal. After inspection the Wartoneers felt that there was a lot of life left in the machines and initiated 1,000-hour inspections, complete overhaul and current stage modifications not previously incorporated. The war-weary status was then removed and all the aircraft went on to units of the 9th Air Force. Another job undertaken by Warton was to commence removing engines from P-51s if they had done over two hundred hours, advisable because of the wear showing on some of the aircraft power plants, and during the hundred-hour inspection and modification programmes replacement engines were fitted as required.

The request for enlisted men to be transferred to the infantry came from B.A.D.A. on 22 January and thirty-three men from the Administrative Division were assigned for front-line duty. Three days later orders came through that Warton's quota of personnel for transferring to the infantry was three hundred and ninety-five, and these men must be fully equipped and ready to leave by 30 January. The situation was obviously acute when it was stated that prisoners could be released from the Guard House if they were minor offenders who volunteered! A greater shock came by teletype from B.A.D.A. Headquarters on 12 February 1945 when it was stated that a second quota of infantry reinforcements was required. This time four hundred and fifty men were needed with regular Army Air Force men exempt unless they wished to volunteer.

On 17 February a delegation arrived at Warton to survey the potentialities of the base as a post-hostilities school centre. There could be an estimated 8,000 troops of all branches on vocational training.

On 19 February 1945 Warton's Flying Control received word that a B-24 had crashed near Burnley. The aircraft had in fact come down near the Cantelaugh Reservoir outside Worsthorne after hitting the moor while flying in thick cloud. Volunteers were called for and B.A.D. 2 responded by sending rescue teams and a recovery vehicle, but little

Above:
The ramp displays the usual variety of aircraft including two A-20s in the code letters of the 416th Bomb Group, which had traded them in for A-26s. Outside Hangar 7 is **Jackpot**, while in the distance can be seen Dispersal Area No. 1, which was the author's vantage point on his trips to Warton.

Below:
Among the many interesting visitors to Warton was the Dominie X7394 **Merlin V** from the Royal Navy's No. 782 Squadron based at Donibristle, Scotland.

Victory in Sight

A C-47 was specially painted for transporting gifts from the Warton personnel to Russian ex-POWs who were accommodated in an army camp near Worthing in Sussex. The Russians, wearing unbadged British uniforms, unload the eagerly awaited cargo at the R.A.F. aerodrome at Ford on 22 February 1945.

The assembly of the gliders took plenty of muscle as the photograph shows. The folding cockpit section is being fitted to a Waco CG-4.

Another Alert Crew section gathers in front of an A-26 for their photograph. The officers in the rear row are Tom Flowers, Tom Boland and Charlie Himes, with Adam Talaki on the front row. Master Sergeant Paul Oberdorf is minding the dog.

could be done on arrival at the scene as five on board were killed instantly, two were critically injured, and four other crew members had various injuries. Three of the crew died in the days immediately afterwards, but the pilot, First Lieutenant Charles Goeking, survived the accident although he did not leave hospital for over two years. The aircraft, 42-50668, coded 6X-M- of the 491st Bomb Group, had left its base at North Pickenham in Norfolk to fly to B.A.D. 1 at Burtonwood, but the machine had overshot that airfield and headed up into north Lancashire before it crashed while descending through cloud for a visual sighting.

One of the most noteworthy projects during February was the contribution by B.A.D. 2 personnel of a portion of their PX ration including sweets, chocolate and cigarettes to Russian servicemen who had been prisoners-of-war working as forced labour for the Germans, but who were now freed by the Allied advance. A large number of them were at a camp at Worthing in Sussex, but as it would be some time before they returned to their homeland the men of Warton felt they should receive some comforts while they were in England. On 22 February a C-47, named *Bear Hug* especially for the occasion and piloted by Jack Knight, was flown down to

Above:
A fine view of the B-17G 43–38190 as the wheels start to lower for landing after a test flight.

Below:
Another view of the B-17, this time in the circuit above the mud flats. The blurred effect is from the sun glinting off the polished metal of the fin.

Above:
The final shot of '190 gives an excellent impression of the B-17 as the aircraft, with flaps yet to be lowered, lines up for a landing on Warton's Runway 26.

Below:
The men of the Engine Repair Department pose for a souvenir photograph when their task was completed. Major Herman Gauss is in the centre.

The World's Greatest Air Depot

A poor, but interesting, photograph of Warton from the air does serve to illustrate the large number of aircraft on the field. The powered aircraft are parked on the concrete while the gliders completely cover the grass areas between the runways.

the R.A.F. Station at Ford. The last eight miles of the journey was made by truck with a total haul of 6,000 packets of cigarettes, an equal amount of chocolate bars, sweets and gum plus several thousand boxes of biscuits and cake. Warton received a translated letter from the leader of the Russian contingent which read: 'May we give you the deepest of thanks for the wonderful gifts and more important the spirit behind them.'

During February the transfer of one hundred and six men of the Supply Division to the ground forces began to cause problems as the shortage of personnel forced the re-scheduling of shifts which almost eliminated night working. The problem was compounded when it was learned that the base would be receiving a large number of gliders for assembly. Word arrived that 1,600 gliders would be received during the following two months, and with each glider being packed into five boxes it meant that the unit would be handling some 8,000 boxes. Because of the transfers it was obvious that the Supply Division alone could not handle such a task and a detachment of men from B.A.D. 1, who had been working on gliders at the A.A.F. Station at Crookham Common in Berkshire, were assigned to Warton to assist in the assembly. As the gliders arrived and were uncrated the boxes were broken down and the timber was stored for packing use. The assembly of the Waco CG-4A gliders commenced on 22 February 1945.

Aircraft engineering programmes which commenced in February were the equipping with propeller anti-icing of all B-24s destined for the 'Back to the States' project, as no

It was a great day when the C-47s arrived to take the gliders away! The V4-coded C-47s belonged to the 442nd Troop Carrier Group.

The Waco CG-4s with tow-lines attached wait to be hooked up to the C-47s for their flight across the Channel to bases in France.

aircraft would be allowed to depart without this system being fitted. Ten aircraft had arrived in the month requiring this modification. Another project was the manufacture of new forward entrance doors for B-17s, but the shortage of hinges received from the United States was slowing down the work. As the doors were needed urgently by the combat groups it was decided that B.A.D. 2 would produce the hinges themselves by casting and then milling them to a finish. The programme was greatly speeded up by the manufacture of these hinges as the door situation had been given a priority status.

The Ordnance Branch devised a method for producing the T-3 chaff bomb which consisted of splitting a one-hundred-pound bomb into two pieces just above the fins, inserting a wooden frame to hold the chaff, and then modifying the nose of the bomb to carry a timer and charge. The bomb proved to be completely successful and fully met the

The World's Greatest Air Depot

A close-up of the nose insignia of one of the C-47s with the code letters showing it belonged to the 436th Troop Carrier Group. Gliders and their crates can be seen in the background with Hangars 31 and 32 in the distance.

The C-47s arriving for a complete overhaul during March 1945 were in a poor state with some of them having flown over 2,000 hours. After the work was completed, B.A.D. 2 received a message from the 302nd Air Transport Wing expressing great satisfaction for a job well done.

A.A.F.'s requirements.

The A.F.S.C. had reported to the United States that the VIIIth Bomber Command had requested the removal of ball turrets from B-24s and felt that it was unnecessary to install such turrets in new aircraft. The suggestion was adopted and the new B-24M Liberators arrived minus these turrets. Twenty-two of these aircraft were assigned to the 15th Air Force operating in the Mediterranean and the aircraft were processed with the modifications for that theatre of operations. Amazingly, the 15th Air Force called for the turrets to be installed and the B.A.D. 2 turret section drew as many ball turrets as available from the repairable warehouse, and made them serviceable and ready for fitment. This placed a great strain on B.A.D. 2 as the work on the aircraft was already scheduled and the turret installation brought extra problems. As always, the Warton engineers sorted out the programme and between 15 and 21 February sixteen of the B-24Ms were completed. With hangar space available only for two of these aircraft at one time, one machine was processed every eight hours which was something of a record.

Oops! This A-26 taxied into a recently constructed drainage ditch, but unfortunately no one had remembered to tell Base Operations.

Reports from the IXth Bomber Command indicated that the A-26 bombing system was vulnerable to the dampness encountered in operations from airfields in France. More rigid and specific inspection of the affected parts was instituted and experimentation on improvements and remedies was carried out. Since the new B-24Ms were also equipped with the same all-electric A-4 bombing system as the Invaders, similar trouble was anticipated and steps were taken to solve the problem before it arose.

The men of the 310th Ferrying Squadron, who were on detached service with the

Victory in Sight

Major Himes heads back to base in the C-47B 43–48770 after a flight to units in East Anglia.

Test pilots Wally Woltemath, left, and Pete Swank are both prominent members of the B.A.D. 2 Association.

Five officers from the Air Inspector's Section arrived in the C-45F 44–47255. The aircraft, photographed from the control tower, was next to a rather nice but unknown C-47.

Maintenance Division, were relieved of their assignment during February and returned to their parent unit, the 31st Air Transport Group of the 302nd Air Transport Wing. The Ferrying Squadron had been transferred to the 31st A.T.G. from the 27th A.T.G. in November 1944 and were based at A.A.F. Station 519 at Grove. With the transfer of the 310th Detachment the Maintenance Division was relieved of the duty of ferrying aircraft and their organisation and the Flight Test Department reverted to its original functions.

On 12 March B.A.D. 2 received another shock, this time in the shape of a teletype requesting an additional quota of six hundred and forty-seven men to be transferred to Infantry Reinforcement Training commencing 26 March and to be completed by 7 May 1945.

In March the Supply Division received fewer than expected glider crates with 1,270 arriving, enough to complete two hundred

The gliders had gone, but there were still two hundred and twenty-six aircraft on the airfield when this shot was taken.

and fifty-four gliders. The majority of these were the CG-4A type, but a small number of the later CG-15s were also received. The glider situation would not disappear as after 2,000 crates – enough for four hundred gliders – had arrived, information was received that a further eight hundred boxes were in the port of Liverpool awaiting delivery to Warton.

The Armament Department's work schedules began to increase in every section and this, coupled with the loss of personnel to the infantry, put a strain on the department. With a revision of the schedules the work was completed. Additional work included the harmonisation of the gun sight cameras on the P-51 Mustangs, and this was the first time the operation had been carried out at B.A.D. 2. The task was accomplished and the aircraft were being delivered to combat units with this work completed, plus a full load of ammunition. Increased output of A-26 Invaders also put a further load on the Armament

A number of P-47s arrived for overhaul during April including 44–33302, 7J-J of the 404th Fighter Group, who were based at Fritzlar, Germany.

Victory in Sight

A milestone was reached in the Engine Repair Department with the completion of the 2,500th Packard Merlin. Checking the details are Colonel Paul Jackson with Major Herman Gauss to his right, while the others are technical representatives of the Packard Corporation.

Colonel Tom Scott assumed command of B.A.D. 2 on 5 May 1945, replacing Colonel Hutchins who became ill. The Air Force tradition was to live on in the Scott family as Tom's son David was an astronaut on the Apollo XV moon mission in 1971. Tom Scott, who retired as a General, was a member of the Association until he passed away in 1989.

Department as it was requested that the eight additional underwing guns should now be installed.

In a special ceremony during March, Soldier's Medals were awarded to ten enlisted men for their part in the rescue operations after the crash of the B-24 at Freckleton on 23 August 1944. At a later ceremony another award of the Soldier's Medal went to Sergeant Stanley C. Begonsky for his brave action in trying to rescue the crew members of the A-26 Invaders which collided on 29 November 1944.

The C-47s arriving at Warton for complete overhauls were in a particularly poor state with a number of them having well over 2,000 hours' flying time. They had been used constantly under all conditions of service with little time for inspection or maintenance. Hangar 4 took on the task with work consisting of engine changes, fuel tanks removed and replaced, electrical systems converted, new autopilots installed, blind landing equipment fitted, and complete sets of de-icer boots for each aircraft. A great amount of metal work was undertaken with repairs to doors and complete new floors plus battle damage repairs. The undercarriages were removed and cleaned and if any parts of the aircraft were difficult to obtain, then they were manufactured. Many had new paint applied and the whole programme was a credit to B.A.D. 2. The 302nd Air Transport Wing expressed great satisfaction with both the operation and the appearance of the aircraft when they were returned to them.

A surprise visit by Major-General J.W. Jones, Air Inspector, Army Air Forces, Washington D.C., was made on 20 March in the 8th Air Force's only B-25 Mitchell. After a short but encouraging visit the General departed by air to Heston, near London. On the following morning, five officers from the Air Inspector's Section, A.A.F., arrived in a C-45 at 1015 hours and after a detailed and thorough inspection they departed at 1545 hours. Colonel Hutchins was informed that they were very satisfied with everything they had seen at Warton.

A public relations exercise took place on the 23rd when the Mayor of Preston, Councillor James E. Gee, christened a new P-51D *Winged Victory* in a ceremony held outside Hangar 4. The exercise was to provide publicity for the film of the same name which

The World's Greatest Air Depot

was showing the following week in both Preston and Blackpool, with a portion of the proceeds going to U.S.A.A.F. and R.A.F. charitable funds.

Probably the only known incident involving a B-24 glider happened when Sergeant Orville Wrosch was flight engineer on a Liberator, slow-timing some rebuilt engines. Everything was going well with the aircraft climbing through 10,000 feet when Wrosch asked the pilots if they were going on oxygen, but he was advised that they would be going back down at any time. He climbed into the top turret to enjoy the view when he heard a change in engine noise. He looked out over the port wing to see that both engines were being shut down and the propellers feathered. Wrosch got back out of the turret to check his instruments as the nose of the Liberator started to go down. However, the aircraft then went completely quiet and a quick look out showed that both engines on the starboard side were also stopped. As the aircraft was falling fast, he knew that without power there would be no electricity to unfeather the props, so he crawled down to the auxiliary power plant and was able to start it. The pilots restarted the engines, but they were below 4,000 feet at that time so the start-up was not a moment too soon. There was some finger

As the war ended many aircraft were flown to Langford Lodge for disposal, including these A-20s and B-17s.

Returning home after a flight to Langford Lodge is this Douglas A-20K, 44–560, with Pete Swank driving.

pointing between the pilots, but it was decided to put the whole thing down to 'a lack of communication'.

With the sad passing of their Commander-in-Chief, President Roosevelt, the personnel of B.A.D. 2 gathered for a memorial service on Sunday, 15 April 1945. The well attended service was not allowed to interrupt the working schedule of the base as only off-duty personnel were able to take part.

The glider programme, which had been temporarily cancelled, was resumed again during April, but with the shortage of men to disassemble the crates one hundred Spanish prisoners-of-war were made available to the Supplies Division to help alleviate the situation.

Squadron Leader Warwick of the Air Ministry in London visited Warton to survey the possibility of the airfield being used as a post-war municipal airport for Preston. After a meeting with Colonel Hutchins he was accompanied on a tour of the airfield by Flight Lieutenant Frank Halliwell, B.A.D. 2's R.A.F. Liaison Officer.

Warton was on the receiving end of another shock on 26 April as B.A.D. 2's Commanding Officer Colonel Don Hutchins suffered a massive stroke while in his quarters during the evening. He was immediately moved to the Station Hospital where it was diagnosed as a cerebral problem which resulted in his left side being completely paralysed. Colonel Hutchins was reported to be noticeably better and on the road to recovery, but as it was obvious that he could not resume his duties, a new Commanding Officer was appointed. Colonel Tom W. Scott arrived on 4 May 1945. He had moved directly from B.A.D. 1 Burtonwood where he had been C.O. since 6 March 1944.

In the Engine Repair Department the month of April had been a significant one for Packard Merlin V-1650 engine production. Production activity was remarkable, and personnel oversaw the 2,500th engine to be successfully overhauled, tested and dispatched to combat units. It was also noteworthy that the high rate of production attained was due to reduction in man hours for each process, as the personnel had succeeded in halving the five hundred hours originally allocated for the production line operation. Morale-boosting information was received from the United States in a report which stated that the B.A.D. 2 engine unit had achieved the greatest all-time production figures for the V-1650 series, easily beating those of other depots.

Early in May information was received that forty C-47 transports would be arriving at Warton for a complete engine change. This programme had to be given priority over the B-24 redeployment project and three hangars were assigned with space made for parking on the rear area ramp. B.A.D.A. Headquarters requested that the aircraft must be ready for delivery by 22 May 1945, but the Maintenance Division set a deadline for two days earlier.

On 7 May 1945 unofficial reports announced the defeat of Germany at 1600 hours. It was later announced that Prime Minister Winston Churchill would call Victory in Europe (V.E.) Day at 1500 hours the following day and King George VI would speak to the nation at 2100 hours. It was decreed that 8 and 9 May would be set aside as national holidays for the celebration of victory.

The A-20 nearing Blackpool with the central pier, tower, station and Winter Gardens all clearly visible.

Chapter 10
V.E. Day – The Run Down Commences

At 1230 hours on 8 May, Colonel Scott spoke to the B.A.D. 2 personnel over the base Tannoy system confirming the official notification of the German surrender. He declared that all except essential work should cease immediately for a period of thirty-five and one half hours during which time passes could be issued for leave of absence for those desiring them. The Colonel also announced that various church services would be held as well as a V.E. Day celebration party. He asked the troops to remember those who had paid the supreme sacrifice as well as others who had been wounded. He then went on to remind them that Japan was still a fierce enemy and that the job must continue with determination and vigour until the victory was finally won. Colonel Scott led the festivities in a V.E. party which was held that night in the Technical Area Mess Hall, and then lit the large bonfire built at the rear of the base headquarters. At last use was made of some of the wood from the glider crates!

The job did continue with the order to replace the C-47 engines still in force, but by 13 May none of these aircraft had arrived at Warton. The man hour reduction on the B-24 programme was not implemented and work continued on the Liberators at the normal rate until the C-47s started to appear. To add to the work scheduling problem, word was received that for the redeployment project a further fifty-nine B-24s were to arrive from Stansted, fifty-three from Langford Lodge and one from Burtonwood. The order also stated that the requirement was for twelve of these aircraft to be delivered each day commencing 23 May.

Although the war against Germany was over there seemed to be little relaxation of Warton's work activities and it was a surprise to some when they were detailed to take part in church parades on the Sunday following the V.E. announcement. These took place in Blackpool, Preston, Lytham and Kirkham while a detachment from B.A.D. 2 joined a large military parade in Lytham. However, the ending of hostilities in Europe did start things moving on the 'going home' front as on the 14th orders came through for a number of men to return home if they had enough points (these were accumulated by factors such as time overseas, marriage, children, or other dependent relatives). On the following Friday, the 18th, the points rating was exercised as the first shipment of personnel was made from B.A.D. 2.

On that day all other personnel were directed to see the film *Two Down and One to Go* which included talks by various Generals

Officers and enlisted men alike shared the great joy of reading the newspaper headline which everyone had been waiting for. After the intense activity and comradeship, one Wartoneer commented, 'It's great to be going home, but I'm sure going to miss this place!'

These one-hundred-and-ten-gallon, U.S.-made drop tanks are being loaded on to the C-47B 43–48884 for delivery to fighter units stationed in Germany.

Warton's contingent proudly marched through the streets of Preston during the Victory Parade on 14 May 1945. More of Warton's troops paraded through other towns.

V.E. Day – The Run Down Commences

Test pilot Wally Woltemath looking sharp before leaving for the long journey home.

Even today Wally looks sharp while wearing the same 1945 jacket at his home in Valley, Nebraska.

The usual 'I was there' photograph of Spence Thwaites is extremely interesting as the code letters PJ on the C-47 behind are unknown to this day. No official listing of codes exists, but there is a possibility that the aircraft belonged to the 27th Air Transport Group.

and other war leaders on the defeat of Germany and Italy, but stressed the difficulties of a similar victory over the ruthless Japanese. Far from boosting morale, the film made some of the men think they would not be returning home soon after all and would be shipped to the Pacific to continue the fight.

Later in the day, Lieutenant Jim Allen of the 310th Ferrying Squadron took off from Warton in an A-20 for the 'milk run' to Langford Lodge, but just twelve minutes after departing the aircraft exploded and fell into the Irish Sea. The crash was witnessed by Lieutenant Bob Pierce who was airborne in a P-51 and the emergency services were immediately alerted. Within minutes the Fleetwood lifeboat was launched and after a search lasting nearly seven hours the boat found only an oil slick and some small pieces of wreckage, but there was no sign of the pilot. The following day air–sea rescue found other pieces of the aircraft including one which confirmed the aircraft's identity. The war may have ended, but lives could still be lost in the

Mustangs at Speke airport, Liverpool, awaiting shipment to the United States in a complete reversal of their arrival in England. The QL- and YC-coded aircraft are F-6s from the 69th Tactical Reconnaissance Group of the 9th Air Force, while the CY machine is a P-51D of the 8th Air Force's 55th Fighter Group.

*Also at Speke was the P-51D **Passion's Playground**, personal aircraft of Captain Herbert G. Marsh, an ace of the 355th Fighter Group.*

This B-24J 'Carpetbagger', serial 42–50607, ran out of runway when landing on 29 June 1945.

performance of duty.

On 23 May Colonel Hutchins was transferred from the Base Hospital to the U.S. Army's 157th General Hospital at Birkenhead in Cheshire for recuperation before returning to the United States. A number of his senior officers visited him before his departure.

On the last day of the month the 2213th Q.M. Truck Company (Aviation) departed for A.A.F. Station 547 at Abbots Ripton in Huntingdonshire. There was a lot of sadness as the unit had served Warton well, but it was the shape of things to come and the first of many movements.

As far as the aviation work was concerned

One section of the flight line Alert Crew pose with their jeeps.

it was business as usual for B.A.D. 2. During May one hundred and twenty B-24s had arrived for the redeployment project with twelve of these being delivered; twenty-four had been made ready for delivery and a further twenty-two had been handed over to Flight Test. The engine change C-47s had eventually started to arrive – twenty-nine had been received and of these eighteen had been sent to Flight Test. Also in May fifty-one P-51 Mustangs were processed and delivered, although this time it was not to combat units but to Speke at Liverpool for their return to the United States.

The rundown of the various specialised sections had commenced as on 31 May orders were received from Base Air Depot Area that the production of sparkplugs must cease immediately. It was instructed that, at the discretion of Colonel Scott, all other types of unnecessary work could be terminated. The glider project was officially ended with completed airframes to be disposed of and all personnel still involved at that time were transferred to the B-24 project. Further to the orders to terminate the glider project a teletype was received from the IXth Troop Carrier Command to request a survey of all completed gliders at Warton, and that as many as possible should be made operational and ready for delivery. The survey revealed that one hundred and three CG-4As and CG-15As could be collected. Early in June a fleet of C-47s arrived to clear the airfield of the gliders, much to the relief of Air Traffic and the Alert Crew who had to operate as normal as possible, but with the gliders taking up a large area of the airfield. The work on the processing of S-1 M-2 bombsights, K-14 gunsights and C-1 autopilots was also run down and the order to pack for the return to the United States was received. The efficiency of the B.A.D. 2 specialists came as a surprise to B.A.D.A. Headquarters when it was found that the movement of these items required four C-47s. The bombsight storage vault was then used for other purposes and the boring guard duty was no longer required.

On Saturday, 2 June 1945, Colonel Scott had the pleasant task of awarding Bronze Star medals to Lieutenants Florence Jacobs, Hilda Nevin and Helen Lieb, three of Warton's nurses who, while on detachment to the combat zone, had treated and helped to evacuate American soldiers while under fire and had worked long hours under hazardous conditions during their medical care. It was an unusual type of ceremony, but a proud one for B.A.D. 2.

During June the redeployment of the Supply Division began with large shipments of stock being dispatched to the Occupational Air Force in Germany with other material being made ready for its return to America. The glider packing cases provided much needed wood when the demand for shipments became heavy, but the shortage of waterproof paper and even nails became a problem. Besides the American requirement, nearly 1,000 crates of components were shipped out to the Royal Air Force to support the U.S. types of aircraft still being operated by the British. Despite the loss of personnel through demobilisation and transfers, morale was high with anticipation of going home, and to expedite the work a large roller conveyor system was set up to accelerate the packing and stencilling time.

Word was received that twenty-two B-24 Liberators which had sought refuge in neutral Sweden during combat operations would be made flyable and returned to Warton for a complete overhaul, and were to be added to the redeployment project. The aircraft duly arrived in the first few days of June and were processed with the others. By mid-June one hundred and thirty-three B-24s had been delivered, six were ready to be delivered, and the test pilots were working hard as fifty-one were in Flight Test. Crews of Air Transport Command had been detailed to ferry the aircraft out at the rate of five per day, and this had commenced on 26 June. An addition to the B-24 project was a single aircraft which was found to be at A. & A.E.E. Boscombe Down, the R.A.F.'s experimental station.

The reduction in personnel throughout the base spread across most sections, but it did mean that the remainder would have to work harder to meet deadlines if an emergency arose. One such came for the Electrical Section as units operating the Douglas A-26 Invader were reporting A.O.Gs. (Aircraft on Ground). Problems had appeared with wing flap gear boxes and the Maintenance and Repair Departments were urgently modifying them, but it was found that five hundred carbon-pile voltage regulators were needed to complete the job. As large numbers of these had already been shipped it was left for the electricians to recover regulators from the scrapped ones in the salvage, but the required number was eventually met by building, overhauling and modifying the various items. The Hydraulic Branch too was busy and, although working with only a skeleton staff, wheels and brakes for the A-26s were processed.

The Engine Repair Department of the Maintenance Division had been actively engaged in engine testing since its establishment, and on 20 June 1945 the last of 6,164 rolled off the assembly line. The Packard Merlin V-1650 was tested and ran perfectly, but instead of going into stock it was decided that it would be a fitting end to the unit if the engine went into service immediately. The following day it was duly installed in a P-51. The department ceased operations on the last day of June. (The above figure does not

V.E. Day – The Run Down Commences

Above:
Parking was always at a premium at Warton, but this is ridiculous! The B-24J 42–51523, which was ex-458th Bomb Group, was being test flown after being prepared for the redeployment project when a failure of the right landing gear caused the accident.

Below:
Another photograph shows the badly damaged Liberator which did not do much good to the parked A-20G 43–9384.

The Engine Test Department signed off on 20 June 1945 with this photograph of the gang. A number of B.A.D. 2 Association members are in the shot including Robert Boehm, Bob Stromberg, the late Wayne Farmer, Leo Wolz, William Ratzburg, Frank Bushardt with the dog, and Stanley Ruggles, who named the war bond P-51 **Mazie R** *for his mother.*

include the radial engines which totalled one hundred and twenty-six.)

A great tribute came to the Maintenance Division when its chief, Colonel Paul B. Jackson, was awarded the Legion of Merit for his unit's outstanding service at B.A.D. 2. He had reorganised many sections, streamlining and simplifying their operations, and had supervised the establishment of a method of 'hands on' training for inexperienced personnel. His leadership was certainly an inspiration to his men and he was responsible for the remarkable increase in production with records being broken in practically every section. At the time of writing ninety-four-year-old Colonel Jackson is still idolised by the members of the B.A.D. 2 Association who served under his command at Warton.

The redeployment schedule for the Maintenance Division was received from B.A.D.A. Headquarters ordering a reduction in production of thirty-eight per cent in June and July and fifty per cent in August.

With the B-24 redeployment project deliveries well underway, flown by Air Transport Command crews, troops from throughout the E.T.O. were arriving at Warton to fly home.

Each Liberator would carry ten passengers in makeshift seats in the bomb bay and other areas.

The rundown of all sections continued apace and on 22 June 1945 the Propeller Governor Section completed its last work on the Curtiss Electric and Hamilton Standard types. By the end of the month virtually every production line had come to an end, and over half of the Production Planning Section had been transferred or were declared surplus to requirements.

The arrival of 'brass' was almost a daily occurrence, but on 15 July it arrived in force when Generals Houghton, Searcy and Thomas flew in for a tour of the site and discussions with Colonel Scott on the establishment of the Centralised Technical School No. 1, which would retrain servicemen for their role in civilian life. The title was later changed to the Warton American Technical School.

The redeployment of personnel from the Maintenance Division did not set back the programme for the month of July as one hundred and forty-five aircraft were processed. The last of the many came out on 27 July 1945 with the C-47A 42–93725 taking the honours, but on Thursday, 2 August 1945, the last aircraft ever to be delivered from B.A.D. 2, a B-24 Liberator, took off at 0900 hours. Perhaps it was the excitement of the occasion, but there appears to be no record of the aircraft's serial number.

The news of the Japanese surrender on 14 August 1945 reached Warton and at 1015 hours the following morning Colonel Scott announced the victory to his troops with a celebration speech and details of religious services and other events, for 'It is a truly great occasion.'

On Monday, 20 August 1945, 2,000 men, women and children plus the personnel of B.A.D. 2 crowded into the Freckleton memorial playground to hear Colonel Scott formally turn over the playground to the representatives of Freckleton Parish Council. It was just three days short of one year since the B-24 had crashed on to the village killing a total of sixty-one people, and it was to those

V.E. Day – The Run Down Commences

Over 2,000 people packed the Freckleton memorial playground on 20 August 1945 to hear Colonel Tom Scott formally hand over the playground to the leader of Freckleton Parish Council, Mr W. Rawsthorne. Here, Colonel Scott unveils the memorial which bears a bronze plaque in remembrance of those who died in the crash of the B-24.

An AT-24, courier version of the B-25 Mitchell, is seen as it taxies from the tower after a V.I.P. visit.

The last of many! The lone Mustang taxies past the control tower before being flown off to Speke for preparation for the long journey back to the U.S.A.

who died that the playground was dedicated. The playground was built by the donations of Warton personnel and over six hundred men had actively taken part in its construction. The large memorial stone bearing an inscription was unveiled by Colonel Scott and, happily, the area is used for its intended purpose even today.

Four days later there was a mixture of joy and tears as the largest contingent of troops, over three hundred, left Warton for Lytham station for the long train journey to Southampton where they embarked on the *Queen Elizabeth* for the voyage to New York. They were given a rousing send-off by the local people and the men marched through Lytham preceded by Warton's own 523rd Army Air Force band under the direction of Chief Warrant Officer Philip Azzolina.

On the 25th Blackpool's *Evening Gazette* published two letters from Colonel Scott, one thanking the Mayor of Blackpool for the town's friendliness and hospitality and stressing Warton's good fortune at being located near such a fine community. The second letter was to the editor of the newspaper thanking him for his fair and factual reporting, and for his support of the Americans.

The end of the month of August saw the official closing of Base Air Depot No. 2. A release date of 1 September 1945 was given to the Station in a letter from B.A.D.A. Headquarters dated 14 August 1945. The Depot and Base Commanders, Depot Adjutant plus a number of other officers would stay on after the official closing date to dispose of all supplies, property and personnel. All remaining staff moved from their own locations to new temporary headquarters on Site 8, which was situated at the Lytham end of the airfield. All U.S.A.A.F. sections and organisations remaining would move to Site 8 before 3 September 1945 with the entire Station being turned over to the Warton American Technical School by the following day. All warehouses had been cleared out and prepared for the Technical School with the bulk of office supplies and equipment being passed to that organisation.

The Technical School had the whole Warton facility at its disposal to provide technical and industrial training, either fundamentals or refresher courses. In all eighteen trades were catered for with the main categories being aeronautics, automotive, heavy construction, plus metal, building, electrical trades and surveying work. The first 2,151 students commenced classes on 16 September 1945 with the aviation students having a B-17 and a B-24 plus other airframe parts and forty aero-engines from Piper Cub engines up to the 2,000hp Pratt & Whitney R-2800. The School Commandant was Brigadier-General Cyrus H. Searcy and the intake reached 4,000 per eight-week term, but on 3 January 1946 it was announced that the school would close on the 11th of that month. The school had been highly successful at training men for their return to civilian life, and those requiring such skills had passed through. The staff were transferred to a technical school in Germany to continue the work for troops still on the Continent. By the second week in February 1946 Warton had been completely vacated by the Americans and the R.A.F., the Station's original owners, were moving back in.

*The last photograph of Warton's uniquely American-type control tower before it was dismantled in the 1950s. During the 1942 to 1945 period the tower controlled more aircraft movements than could ever be imagined in this day and age. (**Paul Francis**).*

Anticlimax is the only way one could describe the almost empty Warton which once was so vibrant and alive, and it is hard to imagine that over eight hundred aircraft crowded on to the field at one time. The last aircraft ever to be processed at B.A.D. 2 was the C-47A 42–93725 which was rolled out on 27 July 1945; however, the honour of being the last processed aircraft to leave B.A.D. 2 went to a B-24 on 2 August 1945. Unfortunately, the serial number is unknown.

HEADQUARTERS
BASE AIR DEPOT No. 2
APO 635 U.S. Army
OFFICE OF THE COMMANDING OFFICER

AAF Station 582
29 August 1945

TO: All Officers and Enlisted Men of Base Air Depot No. 2

    As we are presently engrossed in the final stages of closing down this Depot, I want to commend and congratulate you officers and men still remaining here who have contributed to its outstanding achievements. I regret that many who also had a part have already left and will not read this message.

    Although I have been your commanding officer for a comparatively short period, I still take pride in the fact that I was considered worthy of being placed in command of this vast installation. It has been a privilege and great pleasure to serve with such a splendid assemblage of officers and men. You have at all times been exceptionally loyal to your purpose, your country and your commanding officer. You have cooperated among yourselves in a superior manner in your combined efforts to attain the highest degree of success in all of your achievements.

    If it were not for the fact that this command has been organized and developed for a purpose now accomplished – that of doing its part in winning the war – it would seem lamentable to see it disintegrate. But we have done our part in bringing the great conflict to a victorious conclusion, and for the efficiency and expediency you are now showing in winding up activities here, I further bestow and emphasize my commendation.

    My parting wish, then, is that you will know and believe that I sincerely appreciate all you have done and the way in which you have done it – and that health, happiness, good fortune and all the best will be your heritage.

T.W. SCOTT
Colonel, Air Corps
Commanding

Epilogue

So the Stars and Stripes were lowered at Warton for the last time. The Americans had stayed on after the end of the war, but it is Base Air Depot No. 2 which will live on in the annals of the 8th Air Force, for without the achievements of this Station and the ingenuity of its men, that mighty organisation could not have functioned as successfully as it did.

Most airfields used by the 8th Air Force are now overgrown with their buildings crumbling, but Warton is still active as an important part of British Aerospace. It is still at the forefront of aviation technology and employs some of the world's leading technicians and specialists. Nothing changes!

Base Air Depot No. 2's motto was 'It can be done', and although its personnel did not fly in combat, the task was no less important, and sacrifices were made to ensure that 'it was done'.

Warton as it appears today. The main runways have been extended to 8,000 feet and the large white building to the left of the photograph is an aircraft assembly hall. A number of other buildings have been added, but basically the airfield looks much the same as it did in those far-off days. Long may it continue!

Appendix A

WARTON AIRFIELD – LAYOUT AND LOCATION

PHOTO 1   An R.A.F. reconnaissance photograph of Warton taken on 5 April 1945. The picture was issued to the U.S.A.A.F. for notation.

PHOTO 2   This photograph, shot on 1 July 1974, gives an excellent impression of the airfield in its wartime layout as the print has excluded the runway extensions. It should be noted that B.A.D. 2's Runway 15/33 is now 14/32.

PHOTO 3   The hangars are clearly shown in this photograph of 5 October 1970. Additional buildings are also visible, as are a number of wartime Nissen huts.

PHOTO 4   A fine panoramic view of Warton shows the airfield's proximity to the River Ribble with the main runway now extending beyond the wartime boundaries. In the left foreground of the photograph can be seen the remains of Site 11, while to the bottom right new houses have been built on to the area of Site 12. Further towards the warehouses on the right of the runway are the bases for the Nissen huts which stood on Site 10.

DRAWING 1   The basic layout of Warton during the final stages of B.A.D. 2's residence.

DRAWING 2   A map showing the location of the airfield in relation to the local surroundings.

PHOTO 1

The World's Greatest Air Depot

PHOTO 2

Warton Airfield – Layout and Location

PHOTO 3

The World's Greatest Air Depot

PHOTO 4

Warton Airfield – Layout and Location

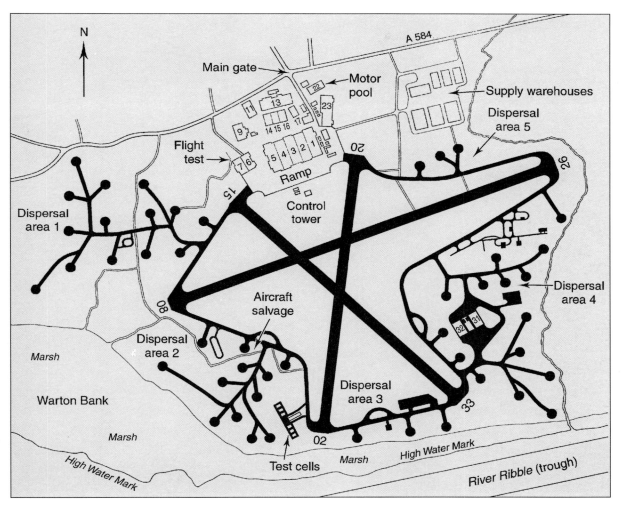

DRAWING 1

The World's Greatest Air Depot

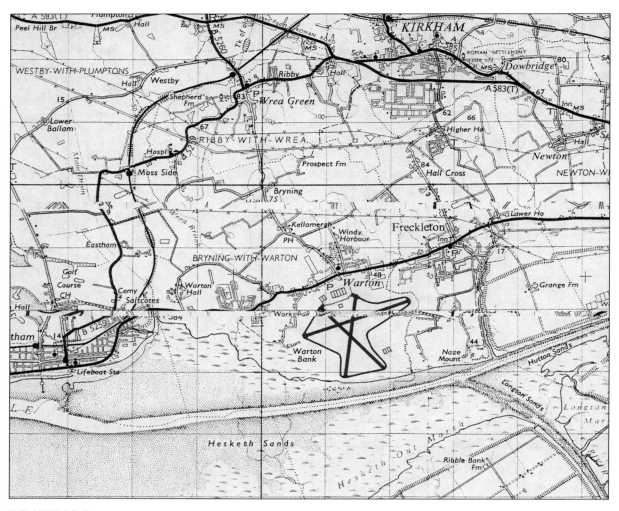

DRAWING 2

Appendix B

WARTON AIRFIELD

Located between Lytham, four miles, and Preston, seven miles, in the county of Lancashire. Selected for U.S. Air Depot in October 1941. Construction work commenced March 1942 with the first U.S.A.A.F. contingent arriving in August 1942. The base was officially transferred to the U.S.A.A.F. on 17 July 1943 and returned to the Royal Air Force in February 1946.
Specifications: Runways – concrete; 08/26 5,631 feet (1,877 yards), 02/20 4,182 feet (1,394 yards), and 15/33 3,960 feet (1,320 yards). Seven main hangars grouped together on north side with two hangars on the south (river) side completed in August 1944. Fifty dispersal stands. Repair shops area – 623,005 square feet. Storage space – 137,363 square feet. Accommodation: ten living sites available for 15,902 persons. Building work: main contractor – Sir Alfred McAlpine & Son Limited with original work by Wimpey Construction Limited.

Organisation

Warton Air Depot was established on 5 September 1942 per General Order No. 19 per Headquarters VIIIth Air Force Service Command. On 21 October 1943 A.A.F. Station 582 became Base Air Depot No. 2 as part of the newly established Base Air Depot Area with headquarters at Burtonwood. B.A.D. 2 was deactivated at midnight on 3 September 1945.

Commanding Officers

Colonel Ira R. Rader	9 October 1942	to	12 February 1943
Colonel A.S. Albro	13 February 1943	to	16 April 1943
Colonel W.D. Lucy	18 April 1943	to	9 May 1943
Colonel C.W. Steinmetz	10 May 1943	to	1 July 1943
Colonel J.J. O'Hara Jr	2 July 1943	to	29 October 1943
Colonel J.G. Moore	30 October 1943	to	24 November 1944
Colonel D.L. Hutchins	25 November 1944	to	4 May 1945
Colonel T.W. Scott	5 May 1945	to	3 September 1945

Appendix C

WARTON AIR DEPOT (B.A.D. 2) COMPLETE OUTPUT

Some 10,068 aircraft were processed and delivered to operational units. This figure does not include almost 4,000 aircraft which were cleared by inspection only and ferried to their respective organisations.

Armament Section

Over 38,430 units had been repaired or overhauled, exclusive of .5 inch (fifty-calibre) machine-guns.

Radio and Signal Section

146,626 units were repaired and made serviceable.

Instrument Section

Repaired and returned to serviceable condition 375,383 items.

Accessories Section

484,800 sparkplugs plus 227,592 other items made serviceable.

Engine Section

Repaired and overhauled 6,164 in-line engines including 3,578 V-1650 Packard Merlins and 2,586 Allison V-1710s. A total of 126 radial engines should be added to the above figure.

Manufacture and Repair

742,800 modification kits were manufactured and made available.

Parachute and Textile Section

Made serviceable were 422,120 items with the breakdown as follows:

Fabrics & Leather	108,350
Life-rafts & vests	52,456
Parachute maintenance	48,692
Flying clothing	108,867
Parachutes repaired	70,184
Electrical clothes	25,307
Miscellaneous	8,264
Total	422,120

Appendix D

SECTION TASKS UNDERTAKEN BY MAINTENANCE DIVISION WARTON AIR DEPOT (B.A.D. 2)

Aircraft Section

Modifications, 100-hour inspections and repair of battle-damaged aircraft. Repairs to new aircraft damaged in transit. Modifications included installation of long-range fuel tanks, water injection systems on engines, installation of bullet-proof glass, armour plating, wing shackles, etc.

Engine Repair

Disassembled and completely overhauled Allison V-1710 and Packard Merlin V-1650 engines, then fully tested them and either installed them into aircraft or shipped them to operational units. Manufactured dies, parts, jigs and fixtures needed to process other equipment.

Manufacturing and Repair

Major work was to manufacture modification kits. Repaired oxygen bottles and engine-driven electrical power units. Repaired propellers, canopies, wings, all flying surfaces. Manufactured and repaired all hand tools plus sheet metal work, welding, painting, heat treatment, plating, carpentry, fabric repairs and renewal, doping and even administrative repairs including typewriters, etc.

Armament

Modified all types of armament equipment, guns, power turrets and bombing equipment. Inspected and made all-echelon repairs of armament, bombsights, computing sights and automatic pilots.

Accessories and Miscellaneous

Completely overhauled and rebuilt all aircraft instruments, cameras, gyros, magnetos, pumps and miscellaneous electrical and hydraulic instruments.

Signal Maintenance and Repair

All radio modifications, repairs and installations plus 100-hour inspections with tune-up of all transmitters and receivers and radio salvage.

Production Control

Recorded and reported all activities of the Maintenance Division and kept the Chief of Maintenance informed on the progress of the modications, repairs and status of aircraft.

Parachute and Textile

Completed repairs on parachutes, electrically-heated flying clothing, life-rafts and Mae Wests. Checked parachutes for deterioration, washed, cleaned and dried them. Modified life-rafts, installed food and water

The World's Greatest Air Depot

kits and fitted sea markers to Mae Wests, etc.

Reclamation and Salvage
Dismantled and salvaged spare parts. The salvageable were repaired and returned to service while the remainder were segregated according to the metal and sold. Salvaged complete aircraft.

Inspection
Checked all newly arrived aircraft and inspected all work completed by the Maintenance Division.

Flight Test
Alert crews pre-flighted all aircraft. Test pilots fully tested aircraft released by the Aircraft Section and any found deficient were returned for additional work. The aircraft cleared were reported to Production Control for the assignment to pilots of the Ferrying Squadrons. The section was also responsible for any visiting aircraft and the billeting of any crew members.

Appendix E

UNITS ASSIGNED TO WARTON

Air Depot

402nd HQ

Airdrome Squadron

10th
45th
100th
361st
363rd
365th

Air Depot Group

7th
10th
16th
23rd
29th
33rd
34th
40th
45th
50th
56th

Air Engineering Squadron

911th

Air Transport Group

27th

Air Transport Squadron

87th
2025th (Provisional)
2920th (Provisional)

Anti-Aircraft Gun Battalion

494th (Mobile)

The World's Greatest Air Depot

Army Air Force Band
523rd

Coastal Artillery Unit
109th Battalion
461st Battalion (Battery B)

Depot Repair Squadron
2nd
7th
10th
16th
22nd
23rd
33rd
35th
45th
61st
65th
89th
91st
307th
310th
311th
314th
315th
316th
319th
325th

Depot Supply Squadron
29th
33rd
312th

Engineer Aviation Battalion
829th
843rd

Ferrying Squadron
310th
311th
312th

Field Hospital
56th

Fire Fighting Platoon-Aviation Engineers
2004th

Headquarters Squadron
29th

Medium Dispatch (Aviation) Squadron
185th to 191st

Military Police Company
976th
977th
1273rd

Ordnance Maintenance Company
315th
2002nd (AF)

Ordnance Medium Maintenance Company
1672nd
1716th

QM Platoon Air Depot Group
407th
433rd

QM Truck Company (Aviation)
407th
1925th
1988th
2212th
2213th
2223rd
2233rd
2461st
2481st
1583rd (Mobile)

The World's Greatest Air Depot

QM Platoon Truck Company

812th (Aviation)

Service Group

1st

Service Squadron

519th
520th

Station Complement Squadron

36th
92nd

Signal Company Service Group

1040th

Station Hospital

307th

Supply Squadron

92nd

Signal Platoon

753rd

Air Base Squadron

16th

Miscellaneous Units

13th Filler Squadron
14th Filler Squadron
15th Filler Squadron
26th Filler Squadron

Appendix F

AIRCRAFT TYPES PROCESSED BY WARTON AIR DEPOT (B.A.D. 2) (1 AUGUST 1943 TO 27 JULY 1945)

Aircraft assembled, repaired or modified, but not including almost 4,000 which passed through Warton for inspection only.

Aircraft	Number
Douglas A-20 Havoc	252
Douglas A-26 Invader	711
Boeing B-17 Fortress	360
Consolidated B-24 Liberator	2,894
Martin B-26 Marauder	1
Beech C-45	2
Douglas C-47 Skytrain	125
Douglas C-53 Skytrooper	3
Noordyne C-64 Norseman	92
Cessna C-78 Bobcat	2
Consolidated C-87 Liberator	3
N. America AT-6 Texan	7
Lockheed A-29	1
Piper L-4 Grasshopper	243
Stinson L-5 Sentinel	67
Waco CG-4 and CG-15 Gliders	387
Lockheed P-38 Lightning	125
Republic P-47 Thunderbolt	338
N. American P-51 Mustang	4,372
Miscellaneous	83
Total	**10,068**

The Warton 'Air Force'

42–85163	AT-6D	*Jet Threat* and *Hoodwink*
41–38607	C-47	*Jackpot*
42–22478	P-47D	*El Champo*
43–6623	P-51B	*Spare Parts*
43–31827	UC-78	*Bamboo Bomber*

310th Ferrying Squadron
41–31742	B-26B	*Demon Deacon*

Radio Call-Signs

Tower	'Farum'
Colonel Jackson	'Jackpot'
Flight Test	'Gorgeous', followed by pilot's name, or phonetic spelling of name initial.

Appendix G

EXAMPLE OF B.A.D. 2 WORK PROCESSING – OCTOBER 1944

P-51	Stage modification	374
F-6	Photo modification	11
	Repair	1
B-24	Stage modification	29
	100-hour inspection	41
	Leaflet modification	2
	Maintenance and Repair	15
P-47	Modification	2
	Repair	3
A-26	Stage modification	8
A-20	Stage modification	26
B-17	Repair	2
UC-64	Assembly	5
C-53	100-hour inspection	2
L-4	Assembly	14
UC-78	Repair	1
P-38	100-hour inspection	6
		Total 542

Appendix H

HANGAR 4 – WORK IN PROGRESS – OCTOBER 1944

Douglas A-26 Invader

41–39257
43–22326
43–22320
43–22291
41–39236
41–39249
43–22329
43–22302
41–39259
41–39263
43–22300
43–22306
43–22322
43–22296
43–22307
41–39233
43–22297
43–22317
43–22301
41–39239
43–22290
43–22324
43–22313
43–22294
43–22292
41–39213
41–39188
41–39219
41–39205
41–39232
41–39210
41–39215
41–39214
41–39227
All above are A-26B models

Consolidated B-24 Liberator

All B-24s are J model except B-24Hs marked *
44–48854
44–40458
44–40441
44–10503
42–52751*

The World's Greatest Air Depot

44–40482
44–48821
44–48864
42–50602
44–48837
44–48903
44–48807
44–48805
44–48851
44–48870
44–48852
44–48820
44–40421
42–95202*
44–48858
44–48850
42–52510*
42–50569
42–51825
44–48806
44–48832
44–48830
44–48829
44–48855
44–48787
44–10597
42–51675
44–48878
42–51289
44–40469
42–51933
42–51666
42–51832
42–51968
42–94797*
41–29594*
42–50585
41–29560*
42–94758*
42–51660
42–51735
42–94811*
42–51170*
42–94802*
42–51936
42–51909
42–51671
42–51939
41–29567*
42–94799*

Appendix I

AIRCRAFT VISITING WARTON

American

Douglas	A-20	Havoc	
Douglas	A-26	Invader	
Lockheed	A-29	Hudson	(British name)
Vultee	A-35	Vengeance	(Tow Target version)

The wintery scene on Warton's ramp shows an interesting visitor in the shape of a Northrop P-61 Black Widow. Only the 422nd and 425th Squadrons operated this large night-fighter and these units were based in France at the time of the visit.

The Beech C-45F was used as a light transport aircraft mainly allocated to headquarters flights for use by staff officers. This aeroplane, 44–47059, appeared to be parked away from the ramp as a snowfall has made the concrete disappear.

A tow-target version of the Marauder was designated AT-23, and 42–95727 carries the code letters of the 4th Gunnery and Tow Target Flight.

The World's Greatest Air Depot

An airliner of the future, this C-54A, 42–72309, would be converted into a civil DC-4 after the war. The aircraft would operate with American Airlines as NC90434. A number of Skymasters still operate in Central and South America even today.

N. American	AT-6	Texan	(British Harvard)
Martin	AT-23	Marauder	(Tow Target version)
N. American	AT-24	Mitchell	(Transport version)
Boeing	B-17	Fortress	
Consolidated	B-24	Liberator	
N. American	B-25	Mitchell	
Martin	B-26	Marauder	
Beech	C-45	Expediter	(British name)
Curtiss	C-46	Commando	
Douglas	C-47	Skytrain	(British Dakota)
Douglas	C-53	Skytrooper	(C-47 type)
Douglas	C-54	Skymaster	
Fairchild	C-61	Forwarder	(British Argus)
Noordyne	C-64	Norseman	(Canadian-built)
Cessna	C-78	Bobcat	
Consolidated	C-87	Liberator	(Transport version)
Waco	CG-4	Glider	(British Hadrian)
Waco	CG-15	Glider	
Lockheed	F-4	Lightning	(Photo recon version of P-38)
Lockheed	F-5	Lightning	(Improved F-4)
N. American	F-6	Mustang	(Photo recon version of P-51)
Piper	L-4	Grasshopper	
Stinson	L-5	Sentinel	
Consolidated	OA-10	Catalina	(U.S.A.A.F. version)
Lockheed	P-38	Lightning	
Bell	P-39	Airacobra	
Republic	P-47	Thunderbolt	
N. American	P-51	Mustang	
Northrop	P-61	Black Widow	

The World's Greatest Air Depot

An interesting visitor in April 1945 was the B-26G 44–68161 of the 394th Bomb Group, who were known as 'The Bridge Busters'. The H9 code belonged to the 586th Squadron.

American Aircraft in British Service
(Except types noted previously)

Curtiss	Tomahawk
Douglas	Boston
Chance Vought	Corsair
Grumman	Avenger
Grumman	Wildcat
Grumman	Hellcat
Lockheed	10 Electra
Lockheed	12
Stinson	Reliant

British

Airspeed	Oxford
Avro	Anson
Avro	Lancaster

The World's Greatest Air Depot

Always the aircraft enthusiast, Jack Knight could not resist photographing this Royal Navy Firefly 1, Z1977, of No. 1771 Squadron stationed at Burscough, just a few miles from Warton.

Blackburn	Botha
Bristol	Beaufighter
De Havilland	Dominie
De Havilland	Mosquito
De Havilland	Tiger Moth
Fairey	Barracuda
Fairey	Firefly
Gloster	Meteor
Handley Page	Halifax
Hawker	Hurricane
Hawker	Typhoon
Miles	Magister
Miles	Master
Percival	Proctor
Short	Stirling
Vickers	Wellington
V-Supermarine	Spitfire
Westland	Lysander

This visiting Hurricane added to Warton's repair quota.

The visit to Warton of one of the Gloster Meteor prototypes created a great deal of excitement among the aviation-minded personnel of B.A.D. 2. The test pilots hung on every word as the R.A.F. pilot described jet flight.

Glossary

A.F.S.C.	Air Force Service Command
A.O.C.	Air Officer Commanding
A.O.G.	Aircraft on Ground
A.T.A.	Air Transport Auxiliary
A.T.C.	Air Traffic Control
A.T.G.	Air Transport Group
A.T.S.	Air Transport Squadron
B.A.D.	Base Air Depot
B.A.D.A.	Base Air Depot Area
C.C.R.C.	Combat Crew Replacement Centre
C.G.	Commanding General
D.G.A.P.(F.)	Director General Aircraft Production (Factories)
E.T.O.	European Theatre of Operations (U.S.)
G. & T.T.	Gunnery and Tow Target
M.A.P.	Ministry of Aircraft Production
Q.M.	Quartermaster
S.A.D.	Strategic Air Depot
T.C.G.	Troop Carrier Group
U.S.A.A.F.	United States Army Air Force

The World's Greatest Air Depot

The World's Greatest Air Depot